Geoenvironmental Mapping: Methods, Theory and Practice

Geoenvironmental Mapping: Methods, Theory and Practice

Contributors

Paulo Cesar Fernandes da Silva, John Canning Cripps et al.

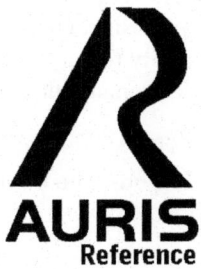

AURIS
Reference

www.aurisreference.com

Geoenvironmental Mapping: Methods,Theory and Practice

Contributors: Paulo Cesar Fernandes da Silva, John Canning Cripps et al.

Published by Auris Reference Limited

www.aurisreference.com

United Kingdom

Geoenvironmental Mapping: Methods,Theory and Practice

ISBN: 978-1-78154-982-7

British Library Cataloguing in Publication Data
A CIP record for this book is available from the British Library

Printed in the United Kingdom

Exclusively distributed by CBS Publishers & Distributors Pvt. Ltd.

Sales & Distribution Rights only for India, Pakistan, Bangladesh, Sri Lanka, Nepal and Bhutan.This book is not to be sold outside these territories.

Contents

List of Abbreviations vii

List of Contributors..ix

Preface.. xiii

Chapter 1 Geo-Environmental Terrain Assessments Based on Remote
Sensing Tools: A Review of Applications to Hazard Mapping
and Control.. 1

Chapter 2 Mineral Prospectivity Mapping Method Integrating Multi-Sources
Geology Spatial Data Sets and Case-Based Reasoning 41

Chapter 3 Mapping Spatial Data on the Web Using Free and Open-Source
Tools: A Prototype Implementation.. 57

Chapter 4 Geostatistical Approach for Site Suitability Mapping of Degraded
Mangrove Forest in the Mahakam Delta, Indonesia 77

Chapter 5 Remote Sensing of Environmental Change in the Antirio Deltaic
Fan Region, Western Greece 99

Chapter 6 Detection and Monitoring of Active Faults in Urban Environments:
Time Series Interferometry on the Cities of Patras and Pyrgos
(Peloponnese, Greece) 115

Chapter 7 Land Use Changes and Environmental Problems Caused by Bank
Erosion: A Case Study of the Kolubara River Basin in Serbia 139

Chapter 8 Environmental Land Use and the Ecological Footprint of Higher
Learning.. 161

Chapter 9 The Role of Tradable Planning Permits in Environmental Land
Use Planning: A Stocktake of the German Discussion..................... 183

Chapter 10 Spatial Analysis for Flood Control by Using Environmental
Modeling.. 201

Chapter 11 Soil Erosion Prediction Using Morgan-Morgan-Finney Model in a
Gis Environment in Northern Ethiopia Catchment........................... 211

Chapter 12   **Geo-Environmental Site Investigation for Municipal Solid Waste Disposal Sites**.. 239

Citations 275

Index.. 277

List of Abbreviations

AHP	Analytic Hierarchy Process
APIs	Application program interfaces
CBR	case-based reasoning
CLR	common language runtime
DEM	Digital elevation model
DEM	Digital elevation model
EROS	Earth Resources Observation and Science
GIS	geographic information systems
GWPs	global warming potentials
IDE	Integrated development environment
IPTA	Interferometric Point Target Analysis
KTF	Kephalonia Transform Fault
MMF	Morgan-Morgan-Finney
ORDBMS	Object-relational database management system
PSI	Persistent Scatterers Interferometry
SLC	Single Look Complex
TM	Thematic Mapper sensorU.S. Geological SurveyUSGS
USLE	Universal Soil Loss Equation
UTM	Universal Transverse MercatorUTM
VES	Vertical electrical soundings
WOE	weights-of-evidence

List of Contributors

Paulo Cesar Fernandes da Silva
Geological Institute - São Paulo State Secretariat of Environment,Brazil

John Canning Cripps
Department of Civil and Structural Engineering, University of Sheffield,United Kingdom

Gebreyesus Brhane Tesfahunegn
College of Agriculture, Aksum University, Aksum, Ethiopia
Centre for Development Research, University of Bonn, Walter-Flex-Street 3, 53113 Bonn, Germany

Lulseged Tamene
International Centre for Tropical Agriculture (CIAT), Chitedze Agricultural Research Station, Lilongwe, Malawi

Paul L. G. Vlek
College of Agriculture, Aksum University, Aksum, Ethiopia

Binbin He
School of Resources and Environment, University of Electronic Science and Technology of China, Chengdu, China

Jianhua Chen
College of Geophysics, Chengdu University of Technology, Chengdu, China

Cuihua Chen
College of Geosciences, Chengdu University of Technology, Chengdu, China

Yue Liu
College of Geosciences, Chengdu University of Technology, Chengdu, China

Sunil Pratap Singh
Department of Physics and Computer Science, Dayalbagh Educational Institute, Agra, India

Preetvanti Singh
Department of Physics and Computer Science, Dayalbagh Educational Institute, Agra, India

Ali Suhardiman
Department of Global Agricultural Sciences, The University of Tokyo, Tokyo, Japan

Satoshi Tsuyuki
Department of Global Agricultural Sciences, The University of Tokyo, Tokyo, Japan

Muhammad Sumaryono
Department of Forest Science, University of Mulawarman, Samarinda, Indonesia

Yohanes Budi Sulistioadi
Department of Forest Science, University of Mulawarman, Samarinda, Indonesia
Division of Geodetic Science, School of Earth Science, The Ohio State University, Columbus, USA

Emmanuel Vassilakis
Department of Dynamics, Tectonics and Applied Geology, Faculty of Geology & Geoenvironment, National & Kapodistrian University of Athens, Panepistimioupoli Zografou, 15784, Athens, Greece

Issaak Parcharidis
Harokopio University of Athens, Department of Geography, El. Venizelou 70, 17671 Athens, Greece

Sotiris Kokkalas
University of Patras, Department of Geology, Division of Physical Geology, Marine Geology and Geodynamics, 265 00 Patras, Greece

Ioannis Fountoulis
National and Kappodistrian University of Athens, Faculty of Geology and Geoenvironment, Department of Dynamic Tectonic and Applied Geology, Panepistimioupolis Zografou, 157 84 Athens, Greece

Michael Foumelis
National and Kappodistrian University of Athens, Faculty of Geology and Geoenvironment, Department of Geophysics and Geothermics, Panepistimioupolis Zografou, 157 84 Athens, Greece

Slavoljub Dragicevic
University of Belgrade, Faculty of Geography, Belgrade, Serbia

Nenad Zivkovic
University of Belgrade, Faculty of Geography, Belgrade, Serbia

Mirjana Roksandic
University of Belgrade, Faculty of Geography, Belgrade, Serbia

Ivan Novkovic
University of Belgrade, Faculty of Geography, Belgrade, Serbia

Stanimir Kostadinov
University of Belgrade, Faculty of Forestry, Belgrade, Serbia

Radislav Tosic
Faculty of Natural Sciences, Banja Luka, Republic of Srpska

Milomir Stepic
Institute for Political Studies,Belgrade, Serbia

Marija Dragicevic
First Elementary School in Obrenovac, Obrenovac, Serbia

Borislava Blagojevic
University of Nis, Faculty of Civil Engineering and Architecture, Serbia

Seth Appiah-Opoku
Geography Department, University of Alabama, Tuscaloosa, USA

Crystal Taylor
Florida State University, USA

Dirk Loehr
Trier University of Applied Sciences, Environmental Campus Birkenfeld, , Germany

Alireza Gharagozlou
Geomatics College of National Cartographic Center of Iran (NCC), Tehran, Iran

Hassan Nazari
Geomatics College of National Cartographic Center of Iran (NCC), Tehran, Iran

Mohammadjavad Seddighi
Geomatics College of National Cartographic Center of Iran (NCC), Tehran, Iran

Giulliana Mondelli,
Institute for Technological Research of São Paulo State, São Paulo State University, University of São Paulo, São Paulo, Brazil

Heraldo Luiz Giacheti and
Institute for Technological Research of São Paulo State, São Paulo State University, University of São Paulo, São Paulo, Brazil

Vagner Roberto Elis
Institute for Technological Research of São Paulo State, São Paulo State University, University of São Paulo, São Paulo, Brazil

Preface

The text *Geoenvironmental Mapping: Methods, Theory and Practice* illustrates the range of environmental geoscience mapping presently carried out around the world. Specialists in several countries have contributed a number of sub disciplinary and thematic topics including landslides, dolines, tsunamis, radon potential, medical geology, rainfall erosion, engineering geology, borehole stratigraphy, lake sediment geochemistry, aggregate resources and remote sensing. In first chapter, a number of case studies are presented to demonstrate the methodological as well as the predictive and preventative aspects of geo-environmental management, with a particular view to regional- and semi-detailed scale, satellite image based terrain classification. Second chapter describes a case-based reasoning (CBR) method for mineral prospectivity mapping that takes spatial features of geology data into account and offers an intelligent approach. Third chapter presents the integration of open-source geospatial tools and web technology to visualize and interact with spatial data using web browser. In fourth chapter, we apply geostatistical and geographical information system (GIS) to generate site-suitability mapping for mangrove rehabilitation guidance in the Mahakam delta. Fifth chapter highlights the important role that analysis of multispectral satellite data can play in the identification of surface alterations related to human activity and natural processes. Sixth chapter concerns the monitoring of the potential ground deformation caused by the active tectonism in the cities of Patras and Pyrgos in Western Greece. A case study of the Kolubara river basin in Serbia has been presented in seventh chapter. Eighth chapter introduces the idea that institutions of higher learning leave ecological footprints on the landscape. Ninth chapter focuses on the role of tradable planning permits in environmental land use planning. Tenth chapter develops a framework for flood control and begins with some general comments on the importance of land use planning and outlines some current environmental issues and then presenting environmental models to use in disaster management plan by using GIS and remote sensing results. The aim of eleventh chapter is to derive spatial distribution of hydrophysical parameters and apply them in the Morgan-Morgan-Finney (MMF) model for estimating soil erosion in the Mai-Negus catchment, northern Ethiopia. Last chapter aims to present and discuss the different tests and steps of a geo-environmental site investigation program proposed for municipal solid waste disposal sites.

Chapter 1

GEO-ENVIRONMENTAL TERRAIN ASSESSMENTS BASED ON REMOTE SENSING TOOLS: A REVIEW OF APPLICATIONS TO HAZARD MAPPING AND CONTROL

Paulo Cesar Fernandes[1] da Silva and John Canning Cripps[2]

[1]Geological Institute - São Paulo State Secretariat of Environment,Brazil

[2]Department of Civil and Structural Engineering, University of Sheffield,United Kingdom

INTRODUCTION

The responses of public authorities to natural or induced geological hazards, such as land instability and flooding, vary according to different factors including frequency of occurrence, severity of damage, magnitude of hazardous processes, awareness, predictability, political willingness and availability of financial and technological resources. The responses will also depend upon whether the hazard is 1) known to be already present thus giving rise to risk situations involving people and/or economic loss; or 2) there is a latent or potential hazard that is not yet present so that development and land uses need to be controlled in order to avoid creating risk situations. In this regard, geo-environmental management can take the form of either planning responses and mid- to long-term public policy based territorial zoning tools, or immediate interventions that may involve a number of approaches including preventative and mitigation works, civil defence actions such as hazard warnings, community preparedness, and implementation of contingency and emergency programmes. In most of cases, regional- and local-scale terrain assessments and classification accompanied by susceptibility and/or hazard maps delineating potential problem areas will be used as practical instruments in efforts to tackle problems and their consequences. In terms of planning, such assessments usually provide advice about the types of development that would be acceptable in certain areas but should be precluded in others. Standards for new construction and the upgrading of existing buildings may also be implemented through legally enforceable building codes based on the risks associated with the particular terrain assessment or classification. The response of public authorities also varies depending upon the information available to make decisions. In some areas sufficient geological information and knowledge about the causes of a hazard may be available to enable an area likely to be susceptible to hazardous processes to be predicted with reasonable certainty. In other places a lack of suitable data may result in considerable uncertainty. In this chapter, a number of case studies are presented to demonstrate the methodological as well as the predictive and preventative aspects of geo-environmental management, with a particular view to regional- and semi-detailed scale, satellite image based terrain classification. If available, information on the geology, geomorphology, covering material characteristics and land uses may be used with remotely sensed data to enhance these terrain classification outputs. In addition, examples provided in this chapter demonstrate the identification and delineation of zones or terrain units in terms of the likelihood and consequences of land instability and flooding hazards in different situations. Further applications of these methods include the ranking of abandoned and/or derelict mined sites and other despoiled areas in support of land reclamation and socio-economic regeneration policies. The discussion extends into policy formulation, implementation of environmental management strategies and enforcement regulations.

USE OF REMOTE DENSING TOOLS FOR TERRAIN ASSESSMENTS AND TERRITORIAL ZONING

Engineering and geo-environmental terrain assessments began to play an important role in the planning process as a consequence of changing demands for larger urban areas and related infra-structure, especially housing, industrial development and the services network. In this regard, the inadequacy of conventional agriculturally-orientated land mapping methods prompted the development of terrain classification systems completely based on the properties and characteristics of the land that provide data useful to engineers and urban planners. Such schemes were then adopted and widely used to provide territorial zoning for general and specific purposes. The process of dividing a country or region into area parcels or zones, is generally called land or terrain classification. Such a scheme is illustrated in Table 1. The zones should possess a certain homogeneity of characteristics, properties, and in some cases, conditions and expected behaviour in response to human activities. What is meant by homogeneous will depend on the purpose of the exercise, but generally each zone will contain a mixture of environmental elements such as rocks, soils, relief, vegetation, and other features. The feasibility and practicability of delineating land areas with similar attributes have been demonstrated throughout the world over a long period of time (e.g. Bowman, 1911; Bourne, 1931; Christian, 1958; Mabbutt, 1968; amongst others), and encompass a wide range of specialisms such as earth, biological and agricultural sciences; hydrology and water resources management; military activities; urban and rural planning; civil engineering; nature and wildlife conservation; and even archaeology. According to Cendrero et al. (1979) and Bennett and Doyle (1997), there are two main approaches to geo-environmental terrain assessments and territorial zoning, as follows. 1) The analytical or parametric approach deals with environmental features or components individually.

Table 1: Hierarchical classification of terrain, soil and ecological units [after Mitchell, 1991]

Terrain unit	Definition	Soil unit	Vegetation unit	Mapping scale (approx.)	Remote sensing platform
Land zone	Major climatic region	Order	-	< 1:50,000,000	
Land division	Gross continental structure	Suborder	Plant panformation to Ecological zone	1:20,000,000 to 1:50,000,000	Meteorological satellites
Land province	Second-order structure or large lithological association	Great group	-	1:20,000,000 to 1:50,000,000	
Land region	Lithological unit or association having undergone comparable geomorphic evolution	Subgroup	Sub-province	1:1,000,000 to 1:5,000,000	Landsat SPOT ERS
Land system *	Recurrent pattern of genetically linked land facets	Family	Ecological region	1: 200,000 to 1:1,000,000	Landsat SPOT, ERS, and small scale aerial photographs
Land catena	Major repetitive component of a land system	Association	Ecological sector	1:80,000 to 1:200,000	
Land facet	Reasonably homogeneous tract of landscape distinct from surrounding areas and containing a practical grouping of land elements	Series	Sub-formation; Ecological station	1:10,000 to 1: 80,000	Medium scale aerial photographs, Landsat, and SPOT in some cases
Land clump	A patterned repetition of two or more land elements too contrasting to be a land facet	Complex	Sub-formation; Ecological station	1:10,000 to 1: 80,000	
Land subfacet	Constituent part of a land facet where the main formative processes give material or form subdivisions	Type	-	Not mapped	Large-scale aerial photographs
Land element	Simplest homogeneous part of the landscape, indivisible in form	Pedon	Ecological station element		

The terrain units usually result from the intersection or cartographic summation of several layers of information [thus expressing the probability limits of findings] and their extent may not corresponding directly with ground features. Examples of the parametric approach for urban planning, hazard mapping and engineering purposes are given by Kiefer (1967), Porcher & Guillope (1979), Alonso Herrero et al. (1990), and Dai et al. (2001). 2) In the synthetic approach, also termed integrated, landscape or physiographic approach, the form and spatial distribution of ground features are analysed in an integrated manner relating recurrent landscape patterns expressed by an interaction of environmental components thus allowing the partitioning of the land into units. Since the advent of airborne and orbital sensors, the integrated analysis is based in the first instance, on the interpretation of remotely sensed images and/or aerial photography. In most cases, the content and spatial boundaries of terrain units would directly correspond with ground features. Assumptions that units possessing similar recurrent landscape patterns may be expected to be similar in character are required for valid predictions to be made by extrapolation from known areas. Thus, terrain classification schemes offer rational means of correlating known and unknown areas so that the ground conditions and potential uses of unknown areas can be reasonably predicted (Finlayson, 1984; Bell, 1993). Examples of the applications of the landscape or physiographic approach include ones given by Christian & Stewart (1952, 1968), Vinogradov et al. (1962), Beckett & Webster (1969); Meijerink (1988), and Miliaresis (2001). Griffiths and Edwards (2001) refer to Land Surface Evaluation as a procedure of providing data relevant to the assessment of the sites of proposed engineering work. The sources of data include remotely sensed data and data acquired by the mapping of geomorphological features. Although originally viewed as a process usually undertaken at the reconnaissance or feasibility stages of projects, the authors point out its utility at the constructional and post-construction stages of certain projects and also that it is commonly applied during the planning of engineering development. They also explain that although more reliance on this methodology for deriving the conceptual or predictive ground model on which engineering design and construction are based, was anticipated in the early 1980s, in fact the use of the methods has been more limited. Geo-environmental terrain assessments and territorial zoning generally involve three main stages (IG/SMA 2003; Fernandes da Silva et al. 2005b, 2010): 1) delimitation of terrain units; 2) characterisation of units (e.g. in bio-geographical, engineering geological or geotechnical terms); and 3) evaluation and classification of units. The delimitation stage consists of dividing the territory into zones according to a set of pre-determined physical and environmental characteristics and properties. Regions, zones or units are regarded as distinguishable entities depending upon their internal homogeneity or the internal interrelationships of their parts. The characterisation stage consists of attributing appropriate properties and characteristics to terrain components. Such properties and characterisitics are designed to reflect the ground conditions relevant to the particular application. The characterisation of the units can be achieved either directly or indirectly, for instance by means of: (a) ground observations and measurements, including in-situ tests (e.g. boring, sampling, infiltration tests etc); (b) laboratory tests (e.g. grain size, strength, porosity, permeability etc); (c) inferences derived from existing correlations between relevant parameters and other data such as those obtained from previous mapping, remote sensing, geophysical surveys and geochemical records. The final stage (evaluation and classification) consists of evaluating and classifying the terrain units in a manner relevant to the purposes of the particular application (e.g. regional planning, transportation, hazard mapping). This is based on the analysis and interpretation of properties and characteristics of terrain - identified as relevant - and their potential effects in terms of ground behaviour, particularly in response to human activities. A key issue to be considered is sourcing suitable data on which to base the characterisation, as in many cases derivation by standard mapping techniques may not be feasible. The large size of areas and lack of accessibility, in particular, may pose major technical, operational, and economic constraints. Furthermore, as indicated by Nedovic-Budic (2000), data collection and integration into useful databases are liable to be costly and time-consuming operations. Such problems are particularly prevalent in developing countries in which suitably trained staff, and scarce organizational resources can inhibit public authorities from properly benefiting from geo-environmental terrain assessment outputs in planning and environmental management instruments. In this regard, consideration has been given to increased reliance on remote sensing tools, particularly satellite imagery. The advantages include: (a) the generation of new data in areas where existing data are sparse, discontinuous or non-existent, and (b) the economical coverage of large areas, availability of a variety of spatial resolutions, relatively frequent and periodic updating of images (Lillesand and Kiefer 2000; Latifovic et al. 2005; Akiwumi and Butler 2008). It has also been proposed that developing countries should ensure that options for using low-cost technology, methods and products that fit their specific needs and capabilities are properly considered (Barton et al. 2002, Câmara and Fonseca 2007). Some examples are provided here to demonstrate the feasibility of a low-cost technique based on the analysis of texture of satellite imagery that can be used for delimitation of terrain units. The delimited units may be further analysed for different purposes such as regional and urban planning, hazard mapping, and land reclamation. The physiographic compartmentalisation technique (Vedovello 1993, 2000) utilises the spatial information contained in images and the principles of convergence of evidence (see Sabins 1987) in a systematic deductive process of image interpretation. The technique evolved from engineering applications of the synthetic land classification approach (e.g. Grant, 1968, 1974, 1975; TRRL 1978), by incorporating and advancing the logic and procedures of geological-geomorphological photo-interpretation (see Guy 1966, Howard 1967, Soares and Fiori 1976), which were then converted to monoscopic imagery (as elucidated by Beaumont and Beaven 1977; Verstappen 1977; Soares et al. 1981; Beaumont, 1985; and others). Image interpretation is performed by identifying and delineating textural zones on images according to properties that take into account coarseness, roughness, direction and regularity of texture elements (Table 2). The key assumption proposed by Vedovello (1993, 2000) is that zones with relatively homogeneous textural characteristics in satellite images (or air-photos) correspond with specific combinations of geo-environmental components (such as bedrock, topography and landforms, soils and covering materials) which share a common tectonic history and land surface evolution. The particular combinations of geo-environmental components are expected to be associated with specific ground responses to engineering and other land-use actions. The process of image interpretation (whether or not supported by additional information) leads to a cartographic product in which textural zones constitute comprehensive terrain units delimited by fixed spatial boundaries The latter correspond with ground features. The units are referred to as physiographic compartments or basic compartmentalisation units (BCUs), which are the smallest units for analysis of geo-environmental components at the chosen cartographic scale (Vedovello and Mattos 1998). The spatial resolution of the satellite image or air-photos being used for the analysis and interpretation is assumed to govern the correlation between image texture and terrain characteristics. This correlation is expressed at different scales and levels of compartmentalisation. Figure 1 presents an example of the identification of basic compartmentalisation units (BCUs) based on textural differences on Landsat TM5 images. In this case the features on images are expressions of differences in the distribution and spatial organisation of textural elements related to drainage network and relief. The example shows the contrast between drainage networks of areas consisting of crystalline rocks with those formed on areas of sedimentary rocks, and the resulting BCUs.

TERRAIN SUSCEPTIBILITY MAPS: APPLICATIONS TO REGIONAL AND URBAN PLANNING

Terrain susceptibility maps are designed to depict ground characteristics (e.g. slope steepness, landforms) and observed and potential geodynamic phenomena, such as erosion, instability and flooding, which may entail hazard and potential damage. These maps are useful for a number of applications including development and land use planning, environmental protection, watershed management as well as in initial stages of hazard mapping applications.

Table 2: Description of elements and properties used for recognition and delineation of distinctive textural zones on satellite imagery [after Vedovello 1993, 2000]

Textural entities and properties	Description
Image texture element	The smallest continuous and uniform surface liable to be distinguishable in terms of shape and dimensions, and likely to be repetitive throughout an image. Usual types of image texture elements taken for analysis include: segments of drainage or relief (e.g. crestlines, slope breaks) and grey tones.
Texture density	The quantity of textural elements occurring within an area on image. Texture density is defined as the inverse of the mean distance between texture elements. Although it reflects a quantitative property, textural density is frequently described in qualitative and relative terms such as high, moderate, low etc. Size of texture elements combined with texture density determine features such as coarseness and roughness.
Textural arrangement	The form (ordered or not) by which textural elements occur and are spatially distributed on an image. Texture elements of similar characteristics may be contiguous thus defining alignments or linear features on the image. The spatial distribution may be repetitive and it is usually expressed by 'patterns' that tend to be recurrent (regularity). For example, forms defined by texture elements due to drainage expressed in rectangular, dendritic, or radial patterns.
Structuring (Degree of spatial organisation)	The greater or lesser organisation underlying the spatial distribution of textural elements and defined by repetition of texture elements within a certain rule of placement. Such organisation is usually expressed in terms of regular or systematic spatial relations, such as length, angularity, asymmetry, and especially prevailing orientations (tropy or directionality). Tropy reflects the anisotropic (existence of one, two, or three preferred directions), or the isotropic (multi-directional or no predominant direction) character of textural features. Asymmetry refers to length and angularity of linear features (rows of contiguous texture elements) in relation to a main feature identified on image. The degree of organisation can also be expressed by qualitative terms such as high, moderate, low, or yet as well- or poorly-defined.
Structuring order	Complexity in the organisation of textural elements, mainly reflecting superposition of image structuring. For example, a regional directional trend of textural elements that can be extremely pervasive, distinctive and superimposed on other orientations also observed on imagery. Another example is drainage networks that display different orders with respect to main stream lines and tributaries (1st, 2nd, 3rd orders)

Early multipurpose and comprehensive terrain susceptibility maps include examples by Dearman & Matula, (1977), Matula (1979), and Matula & Letko (1980). These authors described the application of engineering geology zoning methods to the urban planning process in the former Republic of Czechoslovakia. The studies in this and other countries focused on engineering geology problems related to geomorphology and geodynamic processes, seismicity, hydrogeology, and foundation conditions.

Culshaw and Price (2011) point out that in the UK, a major initiative on urban geology began in the mid-1970s with obtaining geological information relevant to aggregates and other industrial minerals together with investigations relating to the planning of the proposed 3rd London Airport. In the latter case, a very wide range of map types was produced, including one that could be viewed in 3D, using green and red anaglyph spectacles. Of particular interest was the summary "Engineering Planning Map which showed areas that were generally suitable for different types of construction and, also, detailed suggested site investigation procedures (Culshaw and Northmore 2002). As Griffiths and Hearn (2001) explain, subsequently about 50 experimental 'environmental geological mapping, 'thematic'geological mapping' and 'applied geological mapping' projects were carried out between 1980 and 1996 Culshaw and Price (2011) explain that this was to investigate the best means of collecting, collating, interpreting and presenting geological data that would be of direct applicability in land-use planning (Brook and Marker 1987). Maps of a variety of geological and terrain types, including industrially despoiled and potentially unstable areas, with mapping at scales between 1:2500 and 1:25000 were produced. The derivation and potential applications of these sets of maps and reports are described by Culshaw et al. (1990) who explain that they include basic data maps, derived maps and environmental potential maps. Typically such thematic map reports comprise a series of maps showing the bedrock and superficial geology, thickness of superficial deposits, groundwater conditions and areas of mining, fill, compressible, or other forms of potentially unstable ground. Maps showing factual information include the positions of boreholes or the positions of known mine workings. Derived maps include areas in which geological and / or environmental information has been deduced, and therefore is subject to some uncertainty. The thematic sets include planning advice maps showing the constraints on, and potential for, development and mineral extraction. Culshaw et al. (1990) also explained that these thematic maps were intended to assist with the formulation of both local (town or city), regional (metropolis or county) structure plans and policies, provide a context for the consideration of development proposals and facilitate access to relevant geological data by engineers and geologists. It was also recognised that these is a need for national (or state) policies and planning to be properly informed about geological conditions, not least to provide a sound basis for planning legislation and the issuing of advice and circulars. Examples of such advice include planning guidance notes concerning the granting of planning permission for development on potentially unstable land which were published (DOE, 1990, 1995) by the UK government. A further series of reports which were intended to assist planners and promote the consideration of geological information in land-use planning decision making were compiled between 1994 and 1998 by consultants on behalf of the UK government. Griffiths (2001) provides details of a selection of land evaluation techniques and relevant case studies.

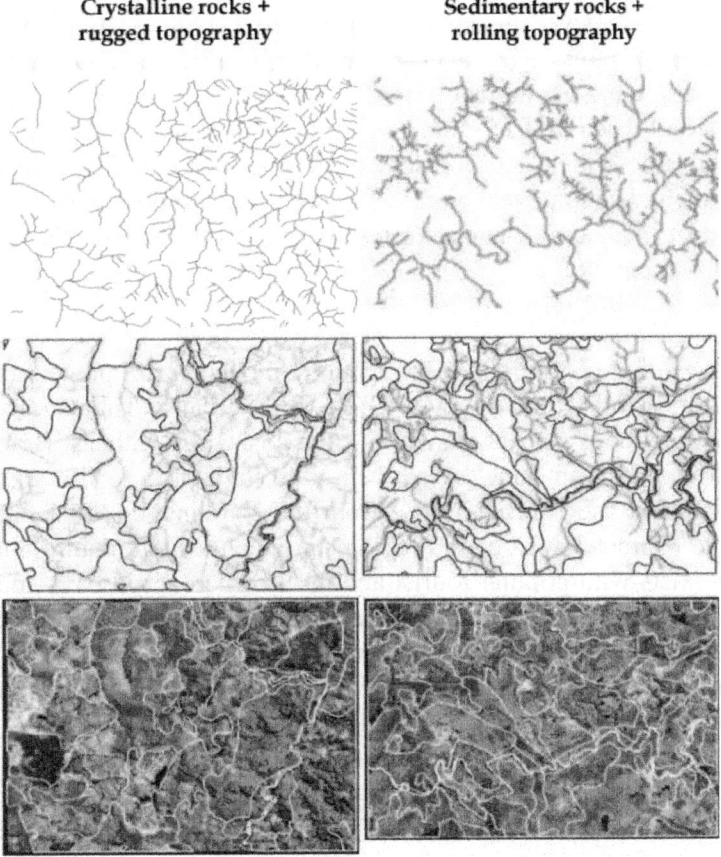

Figure. 1: Identification of basic compartmentalisation units (BCUs) based on textural differences on image. The image for crystalline rocks with rugged topography contrasts with sedimentary rocks with rolling topography. Top: Drainage network. Mid Row: Drainage network and delineated BCUs. Bottom: Composite Landsat TM5 image and delineated BCUs [after Fernandes da Silva et al. 2005b, 2010]

These covered the following themes:

- Environmental Geology in Land Use Planning: Advice for planners and developers (Thompson et al., 1998a)
- Environmental Geology in Land Use Planning: A guide to good practice (Thompson et al., 1998b)
- Environmental Geology in Land Use Planning: Emerging issues (Thompson et al., 1998c)
- Environmental Geology in Land Use Planning: Guide to the sources of earth science information for planning and development (Ellsion and Smith, 1998)

For an extensive review of world-wide examples of geological data outputs intended to assist with urban geology interpretation, land-use planning and utilisation and geological hazard avoidance, reference should be made to Culshaw and Price (2001). Three examples of terrain susceptibility mapping are briefly described and presented in this Section. The physiographic compartmentalisation technique for regional terrain evaluation was explored in these cases, and then terrain units were further characterised in geoenvironmental terms.

Multipurpose Planning

The first example concerns the production of a geohazard prevention map for the City of São Sebastião (IG/SMA 1996), where urban and industrial expansion in the mountainous coastal zone of São Paulo State, Southeast Brazil (Figure 2) led to conflicts in land use as well as to high risks to life and property. Particular land use conflicts arose from the combinations of landscape and economic characteristics of the region, in which a large nature and wildlife park co-exists with popular tourist and leisure encroached bays and beaches, a busy harbour with major oil storage facilities and associated pipelines that cross the area. Physiographic compartmentalisation was utilised to provide a regional terrain classification of the area, and then interpretations were applied in two ways: (i) to provide a territorial zoning based on terrain susceptibility in order to enable mid- to long-term land use planning; and (ii) to identify areas for semi-detailed hazard mapping and risk assessment (Fernandes da Silva et al. 1997a, Vedovello et al., 1997; Cripps et al., 2002). Figure 2 presents the main stages of the study undertaken in response to regional and urban planning needs of local authorities In the Land Susceptibility Map, the units were qualitatively ranked in terms of ground evidence and estimated susceptibility to geodynamic processes including gravitational mass movements, erosion, and flooding. Criteria for terrain unit classification in relation to erosion and mass movements (landslides, creep, slab failure, rock fall, block tilt and glide, mud and debris flow) were the following: a) soil weathering profile (thickness, textural and mineral constituency); b) hillslope profile; c) slope steepness; and d) bedrock structures (fracturing and discontinuities in general). Criteria in relation to flooding included: a) type of sediments; b) slope steepness; and c) hydrography (density and morphology of water courses). The resulting classes of terrain susceptibility can be summarised as follows: Low susceptibility: Areas where mass movements are unlikely. Low restrictions to excavations and man-made cuttings Some units may not be suitable for deep foundations or other engineering works due to possible high soil compressibility and presence of geological structures. In flat areas, such as coastal plains, flooding and river erosion are unlikely. Moderate susceptibility: Areas of moderate to high steep slope (10 to 30%) with little evidence of land instability (small-scale erosional processes may be present) but with potential for occurrence of mass movements. In lowland areas, reported flooding events were associated with the main drainage stream in relevant zones. Terrain units would possess moderate restrictions for land-use with minor engineering solutions and protection measures needed to reduce or avoid potential risks. High susceptibility: Areas of moderate (10 to 20%) and high steep slope (20 to 30%) situated in escarpment and footslope sectors, respectively, with evidence of one or more active land instability phenomena (e.g. erosion + rock falls + landslide) of moderate magnitude. Unfavourable zones for construction work wherein engineering projects would require accurate studies of structural stability, and consequently higher costs. In lowland sectors, recurrent flooding events were reported at intervals of 5 to 10 yrs, associated with main drainage streams and tributaries. Most zones then in use required immediate remedial action including major engineering solutions and protection measures.

Very high susceptibility: Areas of steeper slopes (> 30%) situated in the escarpment and footslope sectors that mainly comprised colluvium and talus deposits. There was evidence of one or more land instability phenomena of significant magnitude requiring full restriction on construction work. In lowland sectors, widespread and frequent flooding events at intervals of less than 5 years were reported and most land-used needed to be avoided in these zones. Units or areas identified as having a moderate to high susceptibility to geodynamic phenomena, and potential conflicts in land use, were selected for detailed engineering geological mapping in a subsequent stage of the study. The outcomes of the further stage of hazard mapping are described and discussed in Section 4.

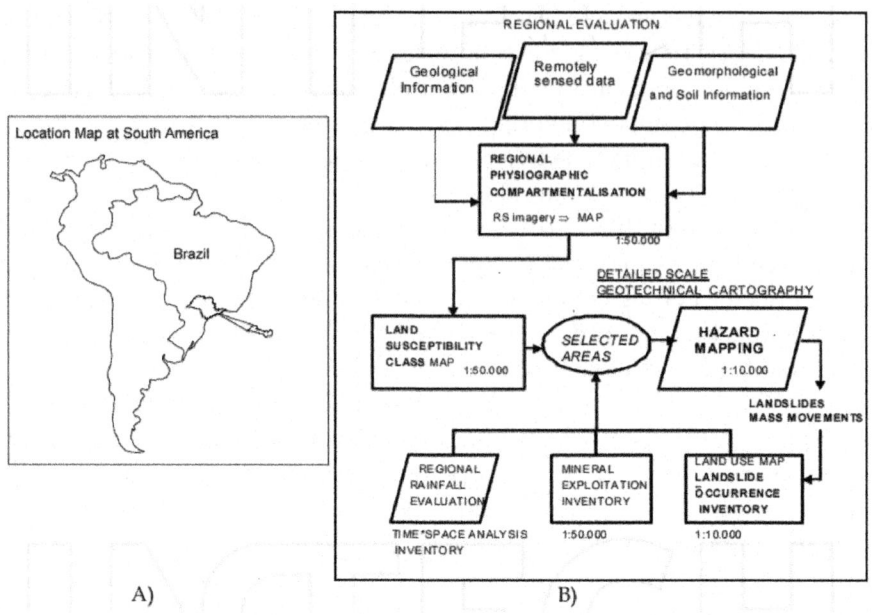

A)   B)

Figure. 2: A) Location map for the City of São Sebastião, north shore of São Paulo State, Southeast Brazil. B) Schematic flow diagram for the derivation of the geohazard prevention chart and structural plan (after IG/SMA, 1996).

Watershed Planning and Waste Disposal

The physiographic compartmentalisation technique was also applied in combination with GIS tools in support of watershed planning in the Metropolitan District of Campinas, central-eastern São Paulo State (Figure 3). This regional screening study was performed at 1:50,000 scale to indicate fragilities, restrictions and potentialities of the area for siting waste disposal facilities (IG/SMA, 1999). A set of common characteristics and properties (also referred to as attributes) facilitated the assessment of each BCU (or terrain unit) in terms of susceptibility to the occurrence of geodynamic phenomena (soil erosion and land instability) and the potential for soil and groundwater contamination. As described by Brollo et al. (2000), the terrain units were mostly derived on the basis of qualitative and semi-quantitative inferences from satellite and air-photo images in conjunction with existing information (maps and well logs – digital and papers records) and field checks. The set of attributes included: (1) bedrock lithology; (2) density of lineaments (surrogate expression of underlying fractures and terrain discontinuities); (3) angular relation between rock structures and hillslope; (4) geometry and shape of hillslope (plan view and profile); (5) soil and covering material: type, thickness, profile; (6) water table depth; and (7) estimated permeability. These attributes were cross-referenced with other specific factors, including hydrogeological (groundwater production, number of wells per unit area), climatic (rainfall, prevailing winds), and socio-political data (land use, environmental restrictions). These data were considered to be significant in terms of the selection of potential sites for waste disposal.

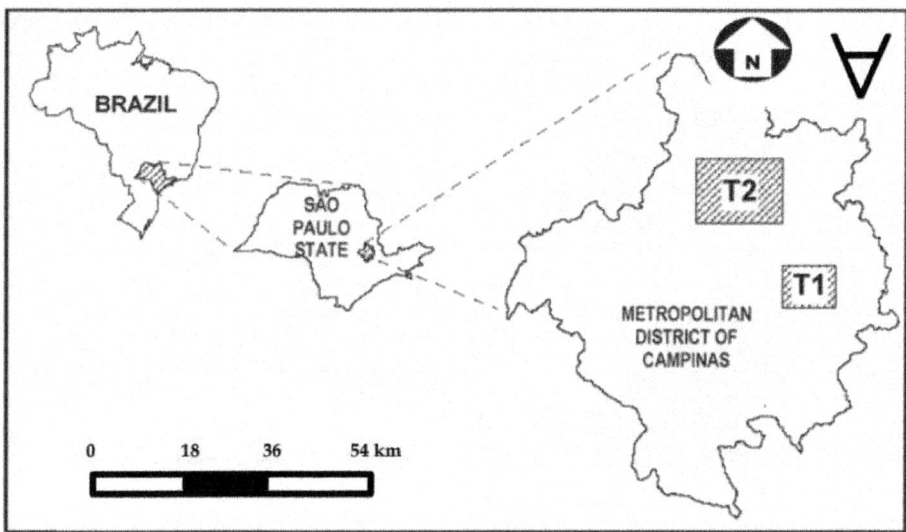

Figure. 3: Location map of the Metropolitan District of Campinas (MDC), central-eastern São Paulo State, Southeast Brazil (see Section 3.2). Detail map depicts Test Areas T1 and T2 within the MDC (see Section 3.3). Scale bar applies to detail map.

Figure 4 displays the study area in detail together with BCUs, and an example of a pop-up window (text box) containing key attribute information, as follows: 1st row - BCU code (COC1), 2nd - bedrock lithology, 3rd - relief (landforms), 4th – textural soil profile constituency, 5th - soil thickness, 6th - water table depth (not show in the example), 7th - bedrock structures in terms of density of fracturing and directionality), 8th - morphometry (degree of dissection of terrain). The BCU coding scheme expresses three levels of compartmentalisation, as follows: 1st letter – major physiographic or landscape domain, 2nd– predominant bedrock lithology, 3rd - predominant landforms, 4th– differential characteristics of the unit such as estimated soil profile and underlying structures. Using the example given in Figure 4, COC1 means: C = crystalline rock basement, O = equigranular gneiss, C = undulating and rolling hills, 1 = estimated soil profile (3 textural horizons and thickness of 5 to 10 m), underlying structures (low to moderate degree of fracturing, multidirectional). In terms of general interpretations for the intended purposes of the study, certain ground characteristics, such as broad valleys filled with alluvial sediments potentially indicate the presence water table level at less than 5 m below ground surface. Flood plains or concave hillside slopes that may indicate convergent surface water flows leading to potentially high susceptibility to erosion, were considered as restrictive factors for the siting of waste disposal facilities (Vedovello et al. 1998).

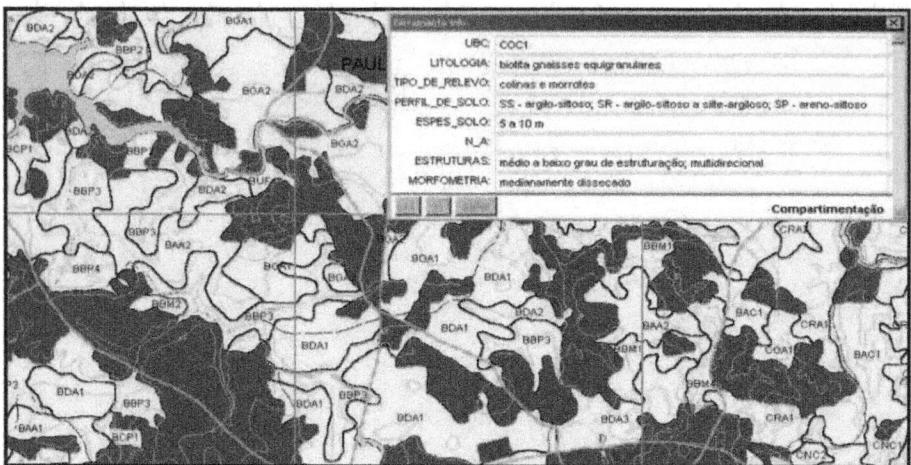

Figure. 4: Basic compartmentalisation units (BCUs) and pop-up window showing key attribute information relevant to BCUs. See text for details. [Not to scale] [after IG/ SMA, 1999]

Regional Development Planning

The third example is a territorial zoning exercise, in which terrain units delimited through physiographic compartmentalisation were further assessed in terms of susceptibility to land instability processes and groundwater vulnerability (Fernandes da Silva et al. 2005b). The study was conducted in two test areas situated in the Metropolitan District of Campinas (Figure 3) in order to assist State of São Paulo authorities in the formulation of regional development policies. It incorporated procedures for inferring the presence and characteristics of underlying geological structures, such as fractures and other discontinuities, then evaluating potential implications to ground stability and the flow of groundwater.

Details of image interpretation procedures for the delimitation of BCUs are described by Fernandes da Silva et al. (2010). The main image properties and image feature characteristics considered were as follows: (a) density of texture elements related to drainage and relief lines; (b) spatial arrangement of drainage and relief lines in terms of form and degree of organisation (direction, regularity and pattern); (c) length of lines and their angular relationships, (d) linearity of mainstream channel and asymmetry of tributaries, (e) density of interfluves, (f) hillside length, and (g) slope forms. These factors were mostly derived by visual interpretation of images, but external ancillary data were also used to assist with the determination of relief-related characteristics, such as slope forms and interfluve dimensions. The example given in Figure 1 shows sub-set images (Landsat TM5) and the basic compartmentalisation units (BCUs) delineated for Test Areas T1 and T2. Based on the principle that image texture correlates with properties and characteristics of the imaged target, deductions can be made about geotechnical-engineering aspects of the terrain (Beaumont and Beaven 1977, Beaumont 1985). The following attributes were firstly considered in the geo-environmental characterisation of BCUs: (a) bedrock lithology and respective weathered materials, (b) tectonic discontinuities (generically referred to as fracturing), (c) soil profile (thickness, texture and mineralogy), (d) slope steepness (as an expression of local topography), and (e) water table depth (estimated). Terrain attributes such as degree of fracturing, bedrock lithology and presence and type of weathered materials were also investigated as indicators of ground properties. For instance, the mineralogy, grain size and fabric of the bedrock and related weathered materials would control properties such as shear strength, pore water suction, infiltration capacity and natural attenuation of contaminants (Vrba and Civita 1994, Hudec 1998, Hill and Rosenbaum 1998, Thornton et al. 2001, Fernandes 2003). Geological structures, such as faults and joints within the rock mass, as well as relict structures in saprolitic soils, are also liable to exert significant influences on shear strength and hydraulic properties of geomaterials (Aydin 2002, Pine and Harrison 2003). In this particular case study, analysis of lineaments extracted from satellite images combined with tectonic modelling underpinned inferences about major and small-scale faults and joints. The approach followed studies by Fernandes and Rudolph (2001) and Fernandes da Silva et al. (2005b) who asserted that empirical models of tectonic history, based on outcrop scale palaeostress regime determinations, can be integrated with lineament analysis to identify areas: i) of greater density and interconnectivity of fractures; and ii) greater probability of open fractures; also to iii) deduce angular relationships between rock structures (strike and dip) and between these and hill slope directions. These procedures facilitated 3-dimensional interpretations and up-scaling from regional up to semi-detailed assessments which were particularly useful for assessments of local ground stability and groundwater flow. The BCUs were then classified into four classes (very high, high, moderate, and low) in terms of susceptibility to land instability and groundwater vulnerability according to qualitative and semi-quantitative rules devised from a mixture of empirical knowledge and statistical approaches. A spreadsheet-based approach that used nominal, interval and numerical average values assigned in attribute tables was used for this. A two-step procedure was adopted to produce the required estimates where, at stage one, selected attributes were analysed and grouped into three score categories (A - high, M - moderate, B - low B) according to their potential influence on groundwater vulnerability and land instability processes. In the second step, all attributes were considered to have the same relative influence and the final classification for each BCU was the sum of the scores A, B, M. The possible combinations of these are illustrated in Table 3. Figure 5 shows overall terrain classifications for susceptibility to land instability.

Table 3: Possible combinations of scores "A" (high), "M" (moderate), and "B" (low) respective to the four attributes (bedrock lithology and weathered materials, fracturing, soil type, and slope steepness) used for classification of units (BCUs) in terms of susceptibility to land instability and groundwater vulnerability.

Combinations of scores	Classification
AAAA	Very high
AAAM, AAAB, AAMM	High
AAMB, AABB, AMMM, AMMB, MMMM	Medium
AMBB, ABBB, MMMB, MMBB, MBBB, BBBB	Low

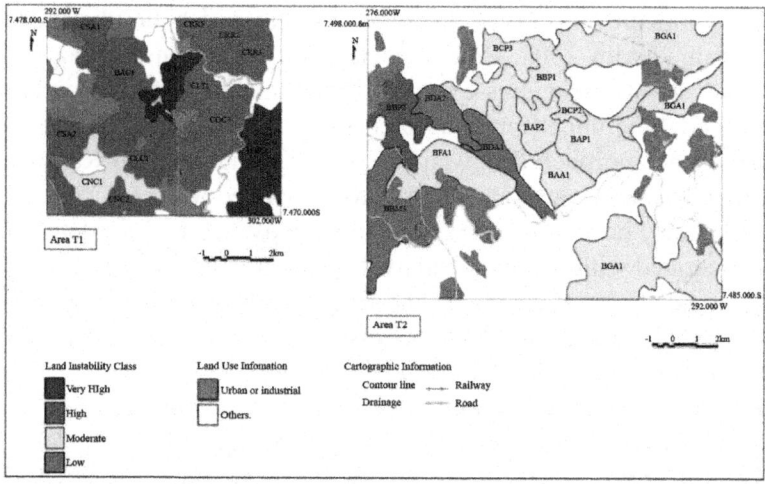

Figure. 5: Maps of susceptibility to land instability processes. Test Areas T1 and T2. UTM projection and coordinates [After Fernandes da Silva et al., 2010]

HAZARD MAPPING: LAND INSTABILITY AND FLOODING

In order to prevent damage to structures and facilities, disruption to production, injury and loss of life, public authorities have a responsibility to assess hazard mitigation and controls that may require remedial engineering work, or emergency and contingency actions. In order to accommodate these different demands, information about the nature of the hazard, and the consequences and likelihood of occurrence, are needed. Hazard maps aim to reduce adverse environmental impacts, prevent disasters, as well as to reconcile conflicting influences on land use. The examples given in this Section demonstrate the identification and zonation in terms of the likelihood and consequences of land instability and flooding hazards. There are several reasons for undertaking such work, for instance to provide public authorities with data on which to base structural plans and building codes as well as civil defence and emergency response programmes.

Application to Local Structural Plans

As indicated in Section 3.1, the BCUs (terrain units) classified as having a moderate to high susceptibility to geodynamic processes (mass movements and flood) were selected for further detailed engineering geological mapping. This was to provide data and supporting information to the structure plan of the City of São Sebastião. The attributes of the selected units were cross-referenced with other data sets, such as regional rainfall distribution, landuse inventory, and mineral exploitation records to estimate the magnitude and frequency of hazards and adverse impacts. Risk assessment was based on the estimated probability of failure occurrence and the potential damage thus caused (security of life, destruction of property, disruption of production). Both the triggering and the predisposing factors were investigated, and, so far as was possible, identified. It is worth noting the great need to consider socio-economic factors in hazard mapping and risk analysis. For instance, areas of consolidated housing and building according to construction patterns and reasonable economic standards were distinguished from areas of unconsolidated/expanding urban occupation. Temporal analysis of imagery and aerial photos, such as densities of vegetation and exposed soil in non-built-up areas, were utilised to supplement the land use inventory. The mineral exploration inventory included the locations of active and abandoned mineral exploitation sites (quarries and open pit mining for aggregates) and certain geotechnical conditions. Besides slope steepness and inappropriate occupancy and land use, the presence of major and minor geological structures was considered to be one of the main predisposing factors to land instability in the region studied. Figure 6 depicts a detail of the hazard map for the City of São Sebastião. Zones of land instability were delimited and identified by code letters that correspond with geodynamic processes as follows: A - landslides, B - creep, C - block tilt/glide, and D - slab failure/rock fall. Within these zones, landsliding and other mass movement hazards were further differentiated according to structural geological predisposing factors as follows: r – occurrence of major tectonic features such as regional faults or brittle-ductile shear zones; f – coincidence of spatial orientations between rock foliation, hillslope, and man-made cuttings; t – high density of fracturing (particularly jointing) in combination with coincidence of spatial orientations between fracture and foliation planes, hillslope, and manmade cuttings (Moura-Fujimoto et al., 1996; Fernandes da Silva et al. 1997b).

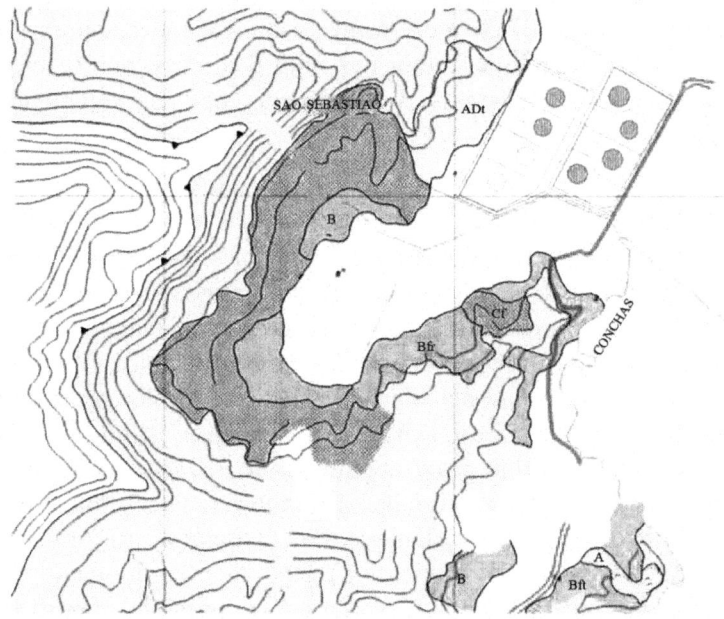

Figure. 6: Example of hazard map from the City of São Sebastião, north shore of São Paulo State, Southeast Brazil. Key for unit classification: Light red = very high susceptibility; Blue = high susceptibility; Light orange = moderate susceptibility; Yellow = low susceptibility. See Section 4.1 for code letters on geodynamic processes and predisposing factors. [after Fernandes da Silva et al. 1997b] (not to scale).

Application to Civil Defence and Emergency Response Programmes

Methods of hazard mapping can be grouped into three main approaches: empirical, probabilistic, and deterministic (Savage et al. 2004, as cited in

Tominaga, 2009b). Empirical approaches are based on terrain characteristics and previous occurrence of geodynamic phenomena in order to estimate both the potential for, and the spatial and temporal distribution of, future phenomena and their effects. Probabilistic approaches employ statistical methods to reduce subjectivity of interpretations. However, the outcomes depend very much on measured patterns defined through site tests and observations, but it is not always feasible to perform this acquisition of data in developing regions and countries. Deterministic approaches focus on mathematical modelling that aims quantitatively to describe certain parameters and rules thought to control physical processes such as slope stability and surface water flow. Their application tends to be restricted to small areas and detailed studies. In the State of Sao Paulo (Southeast Brazil), high rates of population influx and poorly planned land occupation have led to concentration of dwellings in unsuitable areas, thus leading to increasing exposure of the community to risk and impact of hazard events. In addition, over the last 20 years, landsliding and flooding events have been affecting an increasingly large geographical area, so bringing about damage to people and properties (Tominaga et al. 2009a). To deal with this situation, Civil Defence actions including preventive, mitigation, contingency (preparedness), and emergency response programmes have been implemented. The assessment of the potential for the occurrence of landslides, floods and other geodynamic processes, besides the identification and management of associated risks in urban areas has played a key role in Civil Defence programmes. To date, systematic hazard mapping has covered 61 cities in the State of São Paulo, and nine other cities are currently being mapped (Pressinotti et al., 2009). Examples that mix empirical and probabilistic approaches are briefly presented in this Section. The concepts of hazard mapping and risk analysis adopted for these studies followed definitions provided in Varnes (1984) and UN-ISDR (2004), who described risk as an interaction between natural or human induced hazards and vulnerable conditions. According to Tominaga (2009b), a semi-quantitative assessment of risk, R, can be derived from the product R = [H x (V x D)], where: H is the estimated hazard or likelihood of occurrence of a geodynamic and potentially hazardous phenomenon; V is the vulnerability determined by a number of physical, environmental, and socio-economic factors that expose a community and/or facilities to adverse impacts; and D is the potential damage that includes people, properties, and economic activities to be affected. The resulting risk, R, attempts to rate the damage to structures and facilities, injury and loss of lives, and disruption to production. The first example relates to hazard mapping and risk zoning applied to housing urban areas in the City of Diadema (Marchiori-Faria et al. 2006), a densely populated region (around 12,000 inhab. per km2) of only 31.8 km2, situated within the Metropolitan Region of the State Capital – São Paulo (Figure 7). The approach combined the use of high-resolution satellite imagery (Ikonos sensor) and ortho-rectified aerial photographs with ground checks. The aim was to provide civil defence authorities and decision-makers with information about land occupation and ground conditions as well as technical advice on the potential magnitude of instability and flooding, severity of damage, likelihood of hazard, and possible mitigating and remedial measures. Driving factors included the need to produce outcomes in an updateable and reliable manner, and in suitable formats to be conveyed to non-specialists. The outcomes needed to meet preventive and contingency requirements, including terrain accessibility, linear infrastructure conditions (roads and railways in particular), as well as estimations of the number of people who would need to be removed from risk areas and logistics for these actions. Risk zones were firstly identified through field work guided by local authorities. Site observations concentrated on relevant terrain characteristics and ground conditions that included: slope steepness and hillslope geometry, type of slope (natural, cut or fill), soil weathering profile, groundwater and surface water conditions, and land instability features (e.g. erosion rills, landslide scars, river undercutting). In addition, information about periodicity, magnitude, and effects of previous landsliding and flooding events as well as perceptions of potential and future problems were gathered through interviewing of residents. Satellite images were further used to assist with the identification of buildings and houses liable to be affected and the delineation of risk zone boundaries. Risk assessment was based on a qualitative ranking scheme with four levels of risk: R1 (low); R2 (moderate); R3 (high); R4 (very-high). Low risk (R1) zones, for example, comprised only predisposing factors to instability (e.g. informal housing and cuttings in steep slope areas) or to flooding (e.g. informal housing in lowland areas and close to watercourses but no reported flood within the last 5 years). Very-high risk (R4) zones were characterized by significant evidence of land instability (e.g. presence of cracks in soil and walls, subsidence steps, leaning of trees and electricity poles, erosion rills and ravines, landslide scars) or flooding hazards (e.g. flooding height marks on walls, riverbank erosion, proximity of dwellings to river channel, severe floods reported within the last 5 years). The outcomes, including basic and derived data and interpretations, were integrated and then presented on a geo-referenced computational system designed to respond the needs of data displaying and information management of the State of São Paulo Civil Defence authorities (CEDEC). As described by Pressinotti et al. (2007), such system and database, called Map-Risk, includes cartographic data, interpretative maps (risk zoning), imagery, and layers of cadastral information (e.g. urban street network). The system also enabled generation and manipulation of outputs in a varied set of text (reports), tabular (tables), and graphic information including photographic inventories for risk zones. The system was fully conceived and implemented at low cost, utilizing commercial software available that were customized in this visualisation system through target-script programming designed to achieve user functionalities (e.g. ESRI/MapObjects, Delphi, Visual Basic, OCX MapObjects). Examples of delineated risk zones for the City of Diadema and a display of the Map-Risk functionalities are presented in Figure 7. The second example refers to a flooding hazard mapping performed at regional and local scales in the Paraiba do Sul River Watershed, Eastern São Paulo State (Figure 8), in order to provide a rapid and comprehensive understanding of hazard phenomena and their impacts, as well as to enable application of procedures of data integration and mapping in different socio-economic contexts (Andrade et al. 2010). The information was systematised and processed to allow the build-up of a geo-referenced database capable of providing information for both environmental regional planning (economic-ecological zoning) and local scale hazard mapping for civil defence purposes. The regional evaluation covered all the 34 municipalities located in the watershed, and comprised the following stages of work: 1) survey of previous flooding events reported in newspaper and historical archives; 2) data systematisation and consolidation to translate gathered news into useful pieces of technical information; 3) identification of flooding occurrence locations using Google Earth tools; 4) cartographic auditing, geo-referencing and spatial data analysis using a freeware GIS package called SPRING (see Section 5); 5) exploratory statistical analysis of data; 6) preliminary flooding hazard classification on the basis of statistical results. Such preliminary classification used geopolitical (municipality) and hydrographical sub-basin boundaries as units for the analysis.

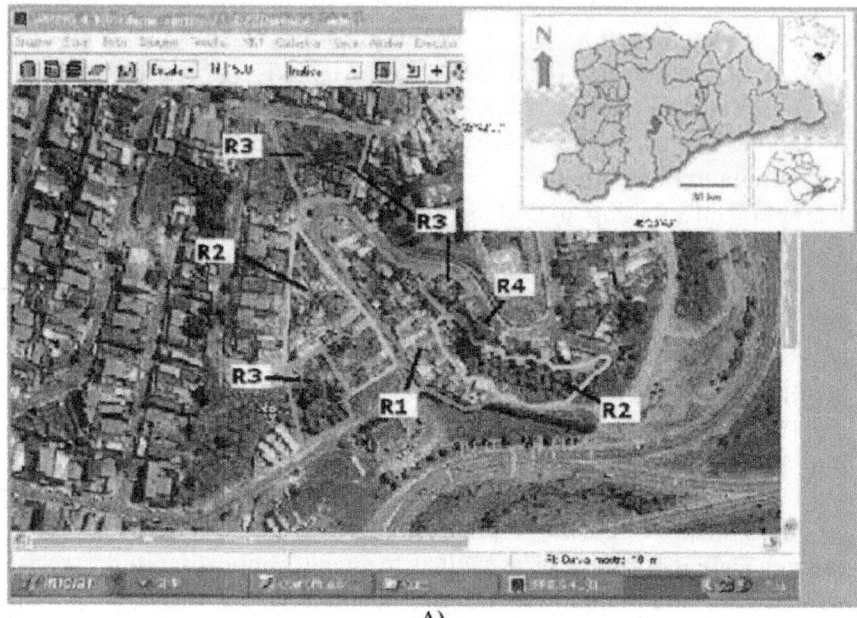

A)

B)

Figure. 7: A) Location of the City of Diadema in the Metropolitan District of São Paulo (State capital), Southeast Brazil and example of delineated risk zones over a high-resolution satellite image (Ikonos). B) Example of Map-Risk system display. See Section 4.2 for details. [after Marchiori-Faria et al., 2006; Pressinotti et al., 2007]

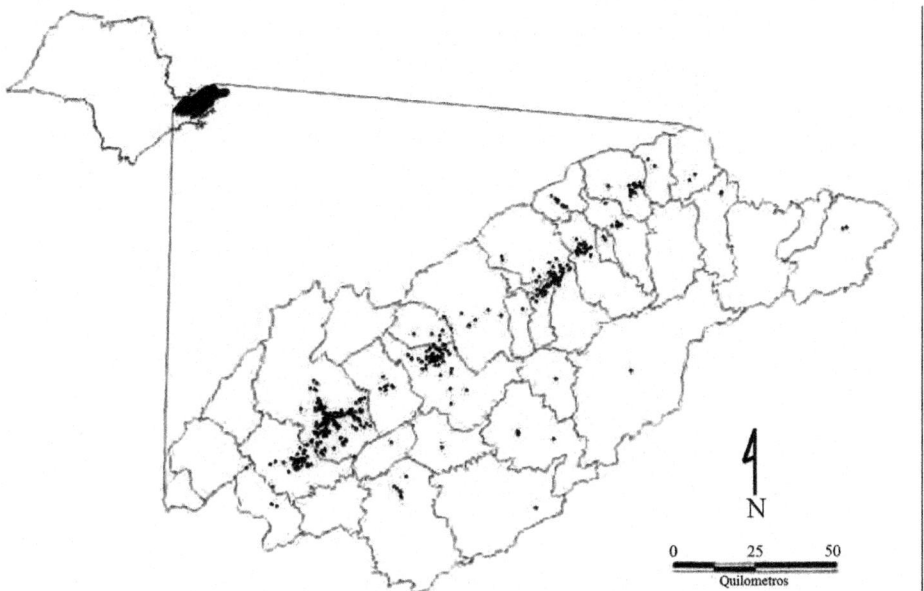

Figure. 8: Location of Paraiba do Sul River Watershed in Eastern São Paulo State and distribution of flooding occurrences. Internal sub-divisions correspond to geopolitical boundaries (municipalities). [After Andrade et al., 2010]

The regional evaluation was followed-up with detailed flooding hazard mapping (1:3,000 scale) in 7 municipalities, which included: a) ground observations - where previous occurrence was reported – to measure and record information on flooding height marks, land occupation, and local terrain, riverbank and water course characteristics; b) georeferencing and spatial data analysis, with generation of interpolated numerical grids on flooding heights and local topography; c) data interpretation and delimitation of flooding hazard zones; d) cross-referencing of hazard zones with land use and economic information leading to delimitation of flooding risk zones. Numerical scoring schemes were devised for ranking hazard and risk zones, thus allowing relative comparisons between different areas. Hazard zone scores were based on intervals of flooding height (observed and interpolated) and temporal recurrence of flooding events. Flooding risk scores were quantified as follows: R = [H x (V x D)], in which potential damage and vulnerability were considered (housing areas, urban infrastructure, facilities and services to be affected) on the basis of image interpretation and cross-referencing with land use maps and information. A detail map (yet unpublished) showing the interpolated grid of flooding heights and delineated hazard zones is presented in Figure 9.

A)

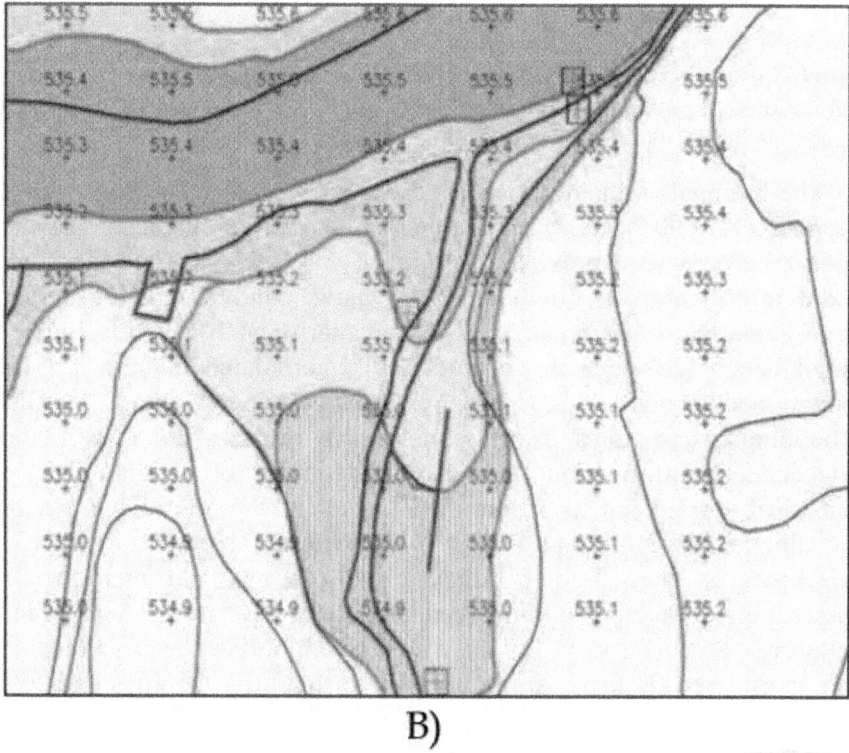

B)

Figure. 9: A) Measurement of maximum flood height for recent flooding event. B) Numerical interpolated grid of flooding heights and delineated flooding hazard zones.

Green = Low probability of occurrence, Estimated flooding heights (Efh) < 0.40 m. Yellow = Moderate probability, 0.40 < Efh < 0.80 m. Light Brown = High probability, 0.80 < Efh < 1.20 m. Red = Very high probability, Efh > 1.20 m. Ground observations and measurements: cross and rectangle. Continuous lines: black = topographic contour lines, blue = main river channel boundaries. Not to scale.

GEO-ENVIRONMENTAL ASSESSMENT: APPLICATIONS TO LAND RECLAMATION POLICIES

Land reclamation of sites of previous mineral exploitation frequently involve actions to minimize environmental damage and aim at re-establish conditions for natural balance and sustainability so reconciling former mined/quarried sites with their surroundings (Brollo et al., 2002). Strategies and programmes for land reclamation need to consider physical and biological characteristics of the local environment as well as socio-economic factors. Socioenvironmental regeneration, involves not only revegetation and land stabilisation engineering, but also rehabilitation or introduction of a new function for the area.

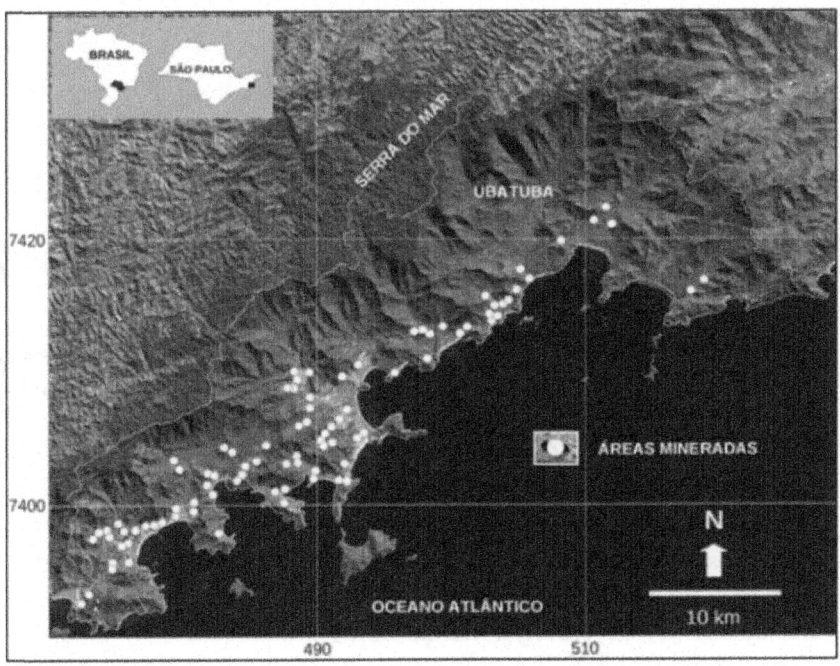

A)

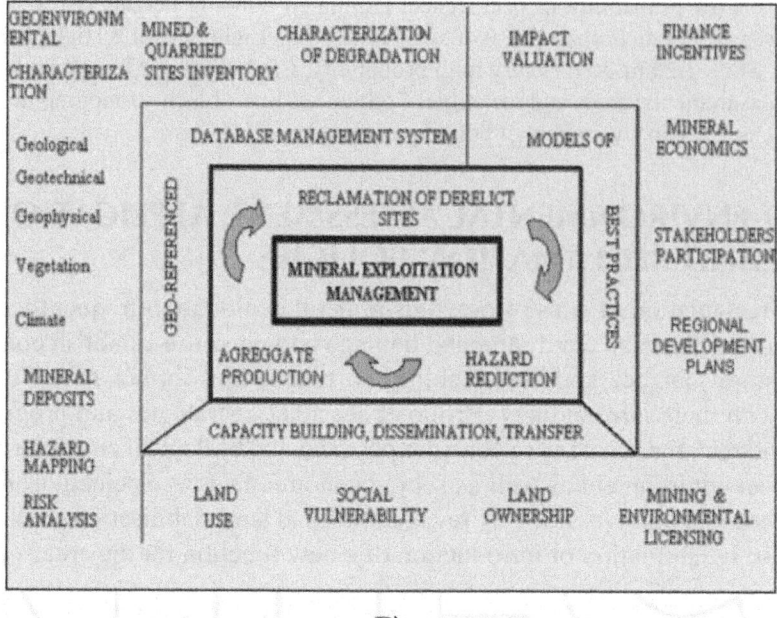

B)

Figure. 10: A) Location map and satellite image of the Municipality of Ubatuba (North Shore São Paulo State, Southeast Brazil). Dots on image represent quarried/mined sites. B) Schematic display of the integrated approach taken to reconcile mineral exploitation management and land reclamation. The scheme shows the three main issues to be addressed (centre) and topics of interest to be studied. [after Ferreira et al., 2006; Ferreira & Fernandes da Silva, 2008]

The case study concerns a GIS-based geo-environmental management scheme to reconcile sustainable mineral exploration of aggregates and construction materials with regeneration of abandoned and/or derelict mined sites in the municipality of Ubatuba (North Shore of São Paulo State, Brazil). Until the early 1990's intensive exploitation of residual soil and ornamental stone (for fill and civil construction) took place in an unplanned and unregulated manner. This led to highly adverse environmental impacts, including the creation of 114 derelict and abandoned sites which resulted in State and Federal authorities enforcing a virtual halt to mining activity in the region. Besides this, the municipality of Ubatuba is highly regarded for its attractive setting and landscape, including encroached coastline with sandy beaches and bays with growing leisure and tourism activities. The area encompasses the Serra do Mar Mountain Range covered by large remnants of Atlantic Forest so that approximately 80% of the municipal territory lies within a nature and wildlife reserve (Figure 10A). As described by Ferreira et al. (2005, 2006), the devised strategy required an integrated approach (Figure 10B) in order to address three key issues: 1) environmental recovery of a number of derelict (abandoned, unsightly) sites; 2) reduction of hazards (land instability, erosion, flooded areas etc), particularly at those sites informally occupied by low income populations; and 3) rational exploitation of materials for local building materials corresponding to local needs. The study was implemented using a freeware GIS and image processing package called SPRING (Câmara et al. 1996, INPE 2009) and ortho-rectified air photos (1-metre resolution, taken in 2001, leading to an approximate scale of 1:3,000). The key output of the land management strategy was a prioritisation ranking scheme based on a comprehensive site critical condition (ICR) score, which synthesised the significance of each factor or issue to addressed (IG/SMA, 2008; Ferreira et al., 2009). Accordingly, the score system consisted of three numerical indicators: 1) environmental degradation indicator (IDE), 2) mineral potential indicator (IPM), and 3) hazard/ risk indicator (IRI). Each indicator was normalised to a scalar range (0 to 1), and the ICR was the sum of the three indicators. The ICR was then used to set up directives and recommendations to advise local and State authorities about the possible measures to be taken, through mid- and long-term policies and/or immediate remedial and mitigating actions. According to Ferreira et al. (2008), the IDE comprised four component criteria to estimate the degree of adverse environmental impact (or degradation) of the individual mineral extraction sites: erosional features, terrain irregularity, herbaceous and bushy vegetation, and exposed soil. Information on these factors was acquired from imagery and ground checks. Tracing of linear features on images was investigated as an indicator of the frequency and distribution of erosional processes (rills, ravines, piping scars) as well as for terrain irregularity. In the first case, the sum of linear features representative of erosional processes was ratioed by the area of each site to quantify the estimate. Similarly, linear features related to the contour of cutting berms, rill marks, and slope breaks caused by mining/quarrying activity, were also measured to quantify terrain irregularity. The areal extent of herbaceous and bushy vegetation as well as exposed soil were also delimited on images. The IPM, as described by Ferreira & Fernandes da Silva (2008), was achieved by means of the following procedures: 1) identification and delimitation of quarried/mined sites (polygons) on geo-referenced imagery; 2) derivation of local DEM (digital elevation model) from topographic contour lines to each delimited site; 3) calculation of local volume to material (V1) based on the original geometry of the quarried/mined sites; 4) calculation of volume of material already taken (V2) and exploitable volume of material (V3), so that [V3] = [V1 − V2]; 5) application of classification rules based on legal environmental and land use restrictions. The calculation of volumes of material (residual soil and ornamental stone) was performed by means of GIS operations involving polygons (areas) and numerical grids of topographic heights generated with nearest neighbour interpolator in the SPRING package (Figure 11). The IRI was derived from R = [H x (V x D)] – see Section 4.2 – focussing on mass movement and flood hazards and their consequences to people, property and economic activity. According to Rossini-Penteado (2007) and Tominaga et al. (2008), Hazard, H, was quantified according to the spatial and temporal probability of occurrence of each phenomenon and then weighted in relation to areal distribution of such probabilities (percentage of sq. km). The vulnerability, V, was computed by means of scores assigned to socio-economic aspects such as nature of built structures, spatial regularity of land occupation, presence of urban infrastructure (e.g. water supply, sanitation, health services, refuse collection and disposal method), road/street network, educational and income patterns. Similarly, in order to estimate the extent of potential damage, D, numerical scores were devised and attributed to the estimated number of people and buildings per unit area, and to the proportion of built area in relation to total area of the site.

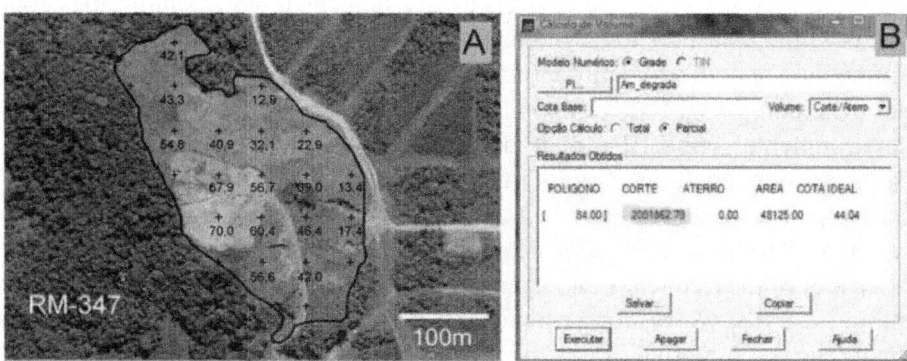

Figure. 11: Example of the mineral potential indicator (IPM) for the Municipality of Ubatuba (North Shore São Paulo State, Southeast Brazil). A) Abandoned/quarried site delimited on image and numerical grid of topographic heights. B) Screen display from GIS-based computation of volumes of exploitable material [after Ferreira & Fernandes da Silva, 2008].

Figure 12 illustrates the application of the ICR scoring scheme to mined/ quarried sites in Ubatuba. In summary, 47% of sites were classified as very low priority, 12% as low priority, 19% as moderate priority, 15% as high priority (18 sites), and 7% as very high priority (8 sites). The priorities represent a combination of availability of exploitable volumes of building materials and the need for measures to tackle adverse environmental impacts and high risk situations (Figure 12). Based on the application of the ICR scores and current land use, directives and recommendations for land reclamation and socio-regeneration of mined/quarried sites were consolidated into ten main groups (IG/SMA, 2008; Ferreira et al., 2009). Such directives and recommendations ranged from simple measures such as routine maintenance and cleaning, revegetation with grass and control of water surface flow (including run-off) to the implementation of leisure and multi-purpose public facilities, major land stabilisation projects combined with mineral exploitation, and monitoring.

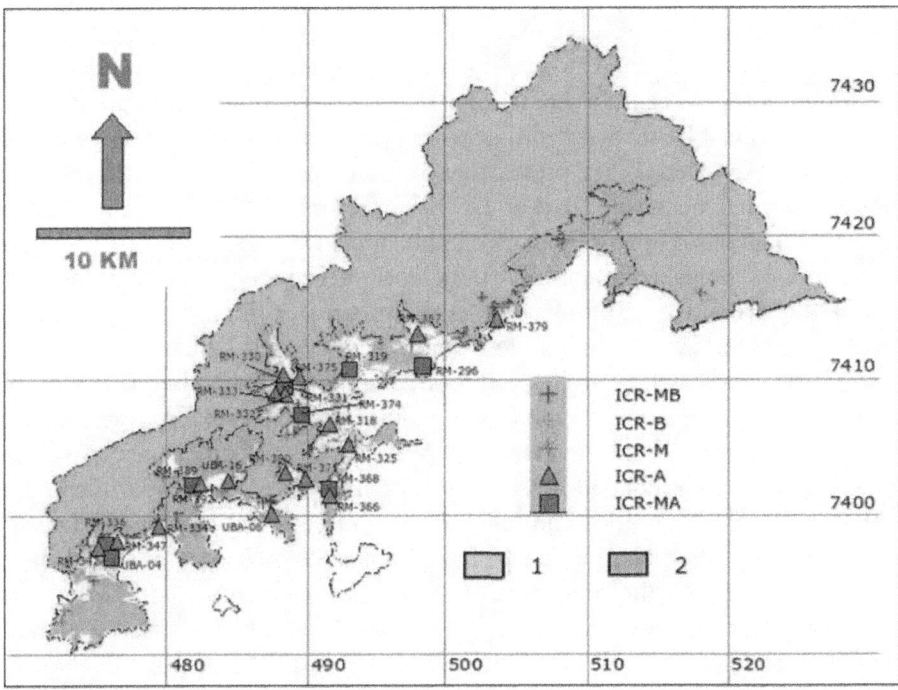

Figure. 12: Spatial distribution of mined/quarried sites classified according to criti-cal condition score scheme (ICR). Sites classified as High (triangles) and Very High (squares) priority are highlighted. Remarks: 1- Serra do Mar Nature and Wildlife State Park, 2- Environmentally sensitive protected areas. ICR-MB = Very Low priority, ICR-B = Low priority, ICR-M = Moderate priority, ICR-A = High priority, ICR-MA = Very high priority [after IG/SMA, 2008].

CONCLUSIONS

Geo-environmental terrain classification may be used as part of the land-use planning decision making and may also provide the basis of responses to emergency situations. In most examples presented here, classification schemes were based on knowledge of the bedrock geology, topography, landforms, superficial geology (soil and weathered materials), groundwater conditions and land-uses. Information for the classification has been variously derived from remote sensing and fieldwork rather than specific site investigations. A framework for carrying out a terrain classification at different scales has been presented. In practice, the effectiveness of land zoning system requires the implementation of planning controls. To do this the Local Authority needs adequate resources and an appropriate legal or planning guidance policy framework. Preferably, the control process should be based on the principle that permission will be given unless there is a good reason for refusal. In granting permission, conditions may then be applied to ensure the safety of the development with regard to landsliding, flooding and other potential problems. However it must be recognised that where practical control over development cannot be exercised, other preventive, mitigative or advisory measures may be all that can be used. Marker (1996) explains that the rate and style of development has a major impact on the information requirements. Without a rigorously enforced planning framework based on accurate information about the ground conditions very rapid urban development will generally lead to construction on areas of less stable land or land which may be subject to hazards such as flooding or pollution. This type of development may also result in the sterilisation of geological resources which it would be expedient to exploit as part of the development process. On the other hand restricting development to designated areas will generally require detailed information about the ground conditions and likelihood and potential impact of hazards to be available at the planning stage. It also assumes existence of the resources and will to enforce the plan. Such models can severely constrain the social and economic development of an area, lead to excessively high population densities and give rise to problems associated with the re-use of previously developed land. Policy formulation may incorporate incentives (e.g. subsidies and reduced taxes) to be provided by local and state governments to encourage such measures and good practice, which can be viewed as kinds of voluntary or induced control. In some cases financial controls exerted by public funding as well as mortgage and insurance providers are the means by which some types of development may be curtailed but such controls may not prevent informal occupancy of hazardous areas. In this regard, some of the examples presented here, from a regional and local perspective, have also demonstrated environmental management regulations may have little meaning in some urban areas subject to rapid expansion where, because of population influx, informal housing is virtually an inevitable consequence. Due to the latter hazard mapping updating and post-episode monitoring [failure episodes] are absolutely vital as these procedures facilitate a contemporary understanding of ground conditions and risk circumstances, which can be essential to provide timely and efficient advice for mitigation and control of hazards as well as to design effective contingency actions and engineering solutions.

ACKNOWLEDGEMENTS

The authors are very much indebted to the Sao Paulo State Geological Institute and its staff for most of the examples provided here.

REFERENCES

1. Akiwumi, F.A., Butler, D.R. (2008). Mining and environmental change in Sierra Leone, West Africa: a remote sensing and hydrogeomorphological study. Environmental Monitoring Assessment, 142: 309–318. ISSN 0167-6369

2. Alonso Herrero, E.; Frances, E., & Cendrero, A. (1990). Environmental geological mapping in Cantabrian Mountains, Spain. In: Proceedings of the 6th Congress of the Intl. Assoc. Engineering Geology and Environment, Amsterdam. Balkema, Rotterdam, 1: 31-35.

3. Andrade, E.; Danna, L.C.; Santos, M.L., & Fernandes da Silva, P.C. (2010). Survey of flooding occurrence in newspaper records as a support for regional planning and hazardmapping. Proceedings of 7th Brazilian Symposium on Geotechnical and Geoenvironmental Cartography. ISSN 2178-1834. Maringa, October 2010. 16p. [in Portuguese]

4. Aydin, F. (2002). Heterogeneity and behaviour of saprolitic slopes. Proceedings of the 9thInternational Congress of the Intl. Assoc. Engineering Geology and Environment, ISBN 0- 620-28559-1, Durban, September 2002. p. 846-856.

5. Barton, J., Alexander, D., Correa, C., Mashelkar, R., Samuels, G., & Thomas S. (2002). Integrating intellectual property rights and development policy. London: UK Department for International Development, Commission on Intellectual Property Rights.

6. Beaumont, T.E. (1985). An application of satellite imagery for highway maintenance andrehabilitation in Niger. International Journal of Remote Sensing, 6 (7): 1263-1267, ISSN 0143-1161.

7. Beaumont, T.E. & Beaven, P.J. (1977). The use of satellite imagery for highway engineering in overseas countries. (T.R.R.L. - Transport & Road Research Laboratory) England, Supplementary Report 279, 19 p.

8. Beckett, P.H.T. & Webster, R. (1969). A Review of Studies on Terrain Evaluation by the Oxford-MEXE-Cambridge Group,1960-1969, MEXE

Report 1123, Christchurch Bell, F.G. (1993). Engineering Geology. Blackwell, Oxford. 269p.

9. Bennett, M.R & Doyle, P. (1997). Environmental Geology. John Wiley and Sons Ltd, Chicester. Bourne, R. (1931). Regional Survey and its Relation to Stocktaking of the Agricultural Resources of the British Empire, Oxford Forestry Memoirs No.13.

10. Bowman, I. (1911). Forest Physiography, Physiography of the US and Principal Soils in Relation to Forestry, John Wiley and Sons, New York.

11. Brollo M.J., Barbosa, J.M., Rocha, F.T., & Martins, S.E. (2002). Research Programme on Characterisation and Reclamation of Environmentally Degraded Areas. Proceedings of the 5th Meeting on Environmental Management Research. São Paulo: CINP/SMA,2002, p. 74-82. [in Portuguese]

12. Brollo, M.J., Vedovello, R., Gutjahr, M.R., Hassuda, S., Iritani, M.A., Fernandes da Silva, P.C., & Holl, MC. (2000). Criteria for selection of areas for waste disposal in regional scale (1:100,000). Application area: Metropolitan Region of Campinas, SaoPaulo State, Brazil. Proceedings of the 8th Congress of the Intl. Assoc. Engineering Geology and Environment. Vancouver, September 1998, 6: 4281-4286.

13. Brook D. & Marker B.R. (1987) Thematic geological mapping as an essential tool in land-use planning. In: Culshaw MG, Bell FG, Cripps JC, O'Hara M (eds) "Planning and Engineering Geology", Engineering Geology Special Publication No. 4. Geological Society, London, pp 211–214.

14. Câmara, G. & Fonseca, F. (2007) Information Policies and Open Source Software in Developing Countries. Journal of the American Society for Information Science and Technology, 58(1): 121–132, ISSN 1532-2890.

15. Câmara, G., Souza, R.C.M., Freitas, U.M., & Garrido, J. (1996). SPRING: Integrating remote sensing and GIS by object-oriented data modelling. Computers & Graphics, 20 (3): 395-403, ISSN 0097-8493.

16. Cendrero, A., Flor, G., Gancedo, R., González-Lastra, J.R., Omenaca, J.S., & Salinas, J.M.(1979). Integrated Assessment and Evaluation of the Coastal Environment of the Province of Vizcaya, Bay of Biscay, Spain. Environmental Geology, 2 (6): 321-331, ISSN 0943-0105.

17. Christian, C.S. (1958). The concepts of land units and land systems. In: Proceedings of the 9th Pacific Science Congress. Vol. 20, pp. 74-81. Christian, C.S. & Stewart, GA. (1952). Summary of general report on Survey of Katherine - Darwin Region. CSIRO Land Research Series No.1

18. Christian, C.S. & Stewart, GA. (1968). Methodology of integrated surveys. Proceedings of theConference on Aerial Surveys and Integrated Studies. UNESCO, Paris. pp. 233-280

19. Cripps, J.C., Fernandes Da Silva, P.C., Culshaw, M.G., Bell, F.G., Maud, R.R., & Foster, A. (2002). The planning response to landslide hazard in São Paulo State - Brazil,

20. Durban - South Africa and Antrim - Northern Ireland. Proceedings of 9thInternational Congress of the Intl. Assoc. Engineering Geology and Environment, ISBN 0- 620-28559-1, Durban, September 2002. p. 1841-1852.

21. Culshaw M.G., Foster, A., Cripps, J.C., & Bell FG (1990). Applied maps for land use planning in Great Britain. Proceedings of the 6th International Congress of the Intl.

22. Assoc. Engineering Geology and Environment. AA Balkema, Rotterdam, 1: 85-93. Culshaw M.G., & Northmore K.J. (2002) An engineering geological map for site investigation planning and construction type identification. In: Van Rooy JL, Jermy

23. CA (eds) Proceedings of the 9th International Association for Engineering Geology and the Environment Congress, Durban, 423–431. South African Institute of Engineering and Environmental Geologists, Pretoria. On CD-ROM.

24. Culshaw M.G. & Price S.J. (2011) The 2010 Hans Cloos lecture: the contribution of urban geology to the development, regeneration and conservation of cities. Bulletin of Engineering Geology and the Environment, 70, Part 3. [In press]

25. Dai, F.C., Lee, C.F., & Zhang, X.H. (2001). GIS-based environmental evaluation for urbanland-use planning: a case study. Engineering Geology, 61 (4): 257 – 271, ISSN 0013- 7952.

26. Dearman, W.R. & Matula, M. (1977). Environmental aspects of engineering geological mapping. Bulletin of the Intl. Assoc. Engineering Geology and Environment, 14: 141-146. ISSN 1435-9529.

27. DOE - Department of the Environment. (1990). Development on unstable ground. Planning Policy Guidance PPG14, HMSO, London.

28. DOE - Department of the Environment. (1995) Development on unstable ground: Landslides and Planning. Planning Policy Guidance PPG14 Annex 1, HMSO, London.

29. Ellison, R. & Smith, A. (1998). Environmental Geology in Land Use Planning: Guide to the sources of earth science information for planning

and development. Published by British Geological Survey (Nottingham, UK) on behalf of Department of Environment, Transport and the Regions, Great Britain.

30. Fernandes, A.J. (2003) The influence of Cenozoic Tectonics on the groundwater vulnerability in fractured rocks: a case study in Sao Paulo, Brazil. Proceedings of the International Conference on Groundwater in Fractured Rocks, Prague, IAH/UNESCO, p.81-82.

31. Fernandes, A.J. & Rudolph, D.L. (2001) The influence of Cenozoic tectonics on the groundwater-production capacity of fractured zones: a case study in Sao Paulo, Brazil. Hydrogeology Journal, 9: 151-167, ISSN 1431-2174.

32. Fernandes Da Silva, P.C., Cripps, J.C., & Wise, S.M. (2005a). The use of Remote Sensing techniques and empirical tectonic models for inference of geological structures: bridging from regional to local scales. Remote Sensing of Environment, 96 (1): 18 – 36. ISSN 0034-4257.

33. Fernandes da Silva, P.C., Ferreira, C.J., Fernandes, A.J., Brollo, M.J., Vedovello, R., Tominaga, L.K., Iritani, M.A., & Cripps, J.C. (2005b)

34. Evaluation of land instability and groundwater pollution hazards applying the physiographic compartmentalisation technique: case study in the Metropolitan District of Campinas, Brazil. Proceedings of the 11th Brazilian Congress on Environmental and Engineering Geology, ISBN 85-7270-017-X. Florianopolis, November 2005. Pp. 383 –402. [In Portuguese]

35. Fernandes da Silva, P.C., Maffra, C.Q.T., Tominaga, L.K., & Vedovello, R. (1997a). Mapping units on São Sebastião Geohazards Prevention Chart, Northshore of São Paulo State, Brazil. In: Environmental Geology, Proceedings of the 30th International Geological Congress. Daoxian, Y. (Ed.), 24: 266-281, VSP Scientific Publisher, ISBN 90-6764-239-8. Utrecht.

36. Fernandes da Silva, P.C., Moura-Fujimoto, N.S.V., Vedovello, R., Holl, M.C., & Maffra, C.Q.T. (1997b). The application of structural geological data in mass movement hazard zoning at the municipality of São Sebastião, Northshore of São Paulo State, Brazil. Proceedings of the 6th Brazilian Symposium on Tectonic Studies, Goiania, May 1997. p. 136-138. [in Portuguese]

37. Fernandes da Silva, P.C., Vedovello, R., Ferreira, C.J., Cripps, J.C., Brollo, M.J., & Fernandes, A.J. (2010) Geo-environmental mapping using physiographic analysis: constraints on the evaluation of land instability and groundwater pollution hazards in the Metropolitan District

of Campinas, Brazil. Environmental Earth Sciences, 61: 1657–1675. ISSN 1866-6280.

38. Ferreira, C.J.; Brollo, M.J.; Ummus, M.E.; & Nery, T.D. (2008). Indicators and quantification of environmental degradation in mined sites, Ubatuba (SP). Revista Brasileira de Geociências, 38(1): 141-152. ISSN 0375-7536 [In Portuguese]

39. Ferreira, C.J. & Fernandes da Silva, P.C. (2008). The use of GIS for prioritising mineral exploitation of building materials in derelict sites: the Ubatuba – SP case. Revista do Instituto Geológico, 29 (1/2): 19-31. ISSN 0100-929X [In Portuguese] Ferreira, C.J., Fernandes da Silva, P.C., & Brollo, M.J. (2009). Directives for socioregeneration of sites affected by mining activity, Ubatuba, Sao Paulo State.

40. Proceedings of the 11th Southeast Brazil Geological Symposia, ISSN 2175-697X, São Pedro, November 2009. pp. 123-124.

41. Ferreira, C.J., Fernandes da Silva, P.C., Brollo, M.J., & Cripps, J.C. (2006). Dereliction problems from exploitation of residual soil and ornamental stone at Sao Paulo State, Brazil. Proceedings of the 10th International Congress of the International Association of Engineering Geology and Environment. Nottingham , September 2006. CD-ROM. 10p.

42. Ferreira, C.J., Fernandes da Silva, P.C, Furlan, S.A., Brollo, M.J., Tominaga, L.K., Vedovello, R., Guedes, A.C.M., Ferreira, D.F., Cripps, J.C., & Perez, F. (2005). Devising strategies for reclamation of derelict sites due to mining of residual soil (Saibro) at Ubatuba, North coast of Sao Paulo State, Brazil: the views and roles of stakeholders. Sociedade e Natureza , Special Issue on Land Degradation, p.643 - 660, ISSN 0103-1570.

43. Finlayson, A. (1984). Land surface evaluation for engineering practices: Applications of the Australian P.U.C.E. system for terrain analysis. Quarterly Journal of Engineering Geology, 17(2): 149-158, ISSN 1470-9236

44. Grant, K. (1968). A Terrain Evaluation System for Engineering. CSIRO, Div. Soil. Mech. Technical Paper No. 2, Melbourne. 27p.

45. Grant, K. (1974). The PUCE Programme for Terrain Evaluation System for Engineering Purposes. II. Procedures for Terrain Classification. CSIRO, Div. Soil. Mech. Technical Paper No. 15, Melbourne. 68p.

46. Grant, K. (1975). The PUCE Programme for Terrain Evaluation System for Engineering Purposes. I. Principles. CSIRO, Div. Soil. Mech. Technical Paper No. 19, Melbourne. 32p.

47. Griffiths, J.S. & Edwards, R.J.G. (2001). The development of land surface

evaluation for engineering practice. In Griffiths, J S (Ed.) Land Surface Evaluation for Engineering Practice. Geological Society, London, Engineering Geological Special Publication, 18, 3 – 9.

48. Griffiths, J S (Ed.) 2001 Land Surface Evaluation for Engineering Practice. Geological Society, London, Engineering Geological Special Publication, 18.

49. Guy, M. (1966). Quelques principles et quelques experiences sur la metodhologie de la photointerpretation. Proceedings of the 2nd International Symposium on Photointerpretation. Paris, 1966. v.1: 2-41.

50. Hill, S.E. & Rosenbaum, M.S. (1998). Assessing the significant factors in a rock weathering system. Quarterly Journal of Engineering Geology, 31: 85-94, ISSN 0013-7952.

51. Howard, A.D. (1967) Drainage analysis in geologic interpretation: a summation. Bulletin of the American Association of Petroleum Geologists, 51: 2246-2254, ISSN 0149-1423.

52. Hudec, P.P. (1998) Rock properties and physical processes of rapid weathering and deterioration. Proceedings of the 8th International Congress of the Intl. Assoc. Engineering Geology and Environment, Vancouver, September 1998. p.335-341.

53. IG/SMA – Instituto Geológico/ São Paulo State Secreariat of Environmet. (1996). São Sebastião Geohazards Prevention Chart, Northshore of São Paulo State, Brazil. São Paulo, Technical Report, 77p. [in Portuguese] IG/SMA - Instituto Geológico/ São Paulo State Secreariat of Environmet. (1999).

54. Methodology for Selection of Suitable Areas for Treatment and Disposal of Solid Waste: Regional Approach. São Paulo, IG/SMA Technical Report, 98p. [in Portuguese]

55. IG/SMA - Instituto Geológico/ São Paulo State Secreariat of Environmet. (2003). Evaluation of land instability and groundwater pollution hazards in the Metropolitan District of Campinas, Brazil. São Paulo, Foreign Commonwealth Office Intl. Cooperation Programme. Technical Report, 57p.

56. IG/SMA - Instituto Geológico/ São Paulo State Secreariat of Environmet. (2008). Directives for Socio-Environmental Regeneration of Derelict Sites Due to Mining of Building Materials, Ubatuba, Northshore of São Paulo State, Brazil. São Paulo, FAPESP – Public Policies Research Programme, Funding Grant n. 03/07182-5. Final Tech. Sci. Report, 168p. [in Portuguese]

57. INPE - Brazilian National Institute for Space Research (2009) SPRING

5.0. http://www.dpi.inpe.br/spring/english/download.php. Cited in 11 December 2009. Kiefer, R.W. (1967). Terrain analysis for metropolitan fringe area planning, Journal of the Urban Planning and Development Division, Proceedings of the American Society of Civil Engineers, UP4, 93, 119-39.

58. Latifovic R., Fytas K., Chen J., & Paraszczak J. (2005) Assessing land cover change resulting from large surface mining development. International Journal of Applied Earth Observation and Geoinformation, 7: 29–48, ISSN 0303-2434.

59. Lillesand, T. & Kiefer, R.W. (2000) Remote Sensing and Image Interpretation. 4th Ed., John Wiley, New York. Mabbut, J.A. 1968. Review of concepts on land classification. In: Land Evaluation. Stewart, G.A. (ed.), p. 11-28, Macmillan, Melbourne.

60. Marchiori-Faria, D.G., Ferreira, C.J., Fernandes da Silva, P.C., Rossini-Pentado, D., & Cripps, J.C. (2006). Hazard mapping as part of Civil Defence preventative and contingency actions: A case study from Diadema, Brazil. Proceedings of the 10th International Congress of the International Association of Engineering Geology and Environment. Nottingham , September 2006. CD-ROM. 10p.

61. Marker, B.R. (1996). Role of the earth sciences in addressing urban resources and constraints In: McCall, G J H ; de Mulder, E F J and Marker, B R (Eds). Urban Geoscience. AGID Special Publication Series No 20 in association with Cogeoenvironment. Balkema (Rotterdam), pp 163-180.

62. Matula, M. (1979). Regional engineering geological evaluation for planning purposes. Bulletin of the Intl. Assoc. Engineering Geology and Environment, 18: 18-24. ISSN 1435-9529

63. Matula, M. & Letko, V. (1980). Engineering geology in planning the metropolitan region of Bratislava. IAEG Bull.,. Bulletin of the Intl. Assoc. Engineering Geology and Environment, 19: 47-52. ISSN 1435-9529

64. Meijerink, A.M. (1988). Data acquisition and data capture through terrain mapping units. ITC Journal, International Journal of Applied Earth Observation and Geoinformation 1: 23-44. ISSN 03032434

65. Miliaresis, G.Ch. (2001). Geomorphometric mapping of Zagros Range at regional scale. Computer and Geosciences, 27 (7): 775 – 786. ISSN 0098-3004. Mitchell, C.W. (1991). Terrain Evaluation. 2nd Ed. Longman, Essex. 497p.

66. Moura-Fujimoto, N.S.V., Holl, M.C., Vedovello, R., Fernandes da Silva, P.C, & Maffra, C.Q.T. (1996). The identification of mass movement

hazard zones at the municipality of São Sebastião, Northshore of São Paulo State, Brazil. Proceedings of the 2nd Brazilian Symposium on Geotechnical Cartography. São Paulo, November 1996. p. 129-137. [in Portuguese]

67. Nedovic-Budic, Z. (2000). Geographic Information Science Implications for Urban and Regional Planning. URISA Journal, 12(2): 81-93. ISSN 1045-8077.

68. Pine, R.J. & Harrison, J.P. (2003). Rock mass properties for engineering design. Quarterly Journal of Engineering Geology and Hydrogeology, 36: 5–16. ISSN1470-9236.

69. Porcher, M. & Guillope, P. (1979). Cartographie des risques ZERMOS appliquées a des plans d'occupation des sols en Normandie. Bull. Mason Lab. des Ponts et Chausses, 99.

70. Pressinotti, M.M.N., Guedes, A.C.M, Fernandes da Silva, P.C, Sultanum, H.J., & Guimarães, R.G. (2007). Automated system for displaying of gravitational mass movement and flooding hazard mapping of São Paulo State. Proceedings of the 6th Brazilian Symposium on Geotechnical and Geo-environmental Cartography. Uberlândia, November 2007. p. 324 – 333. [in Portuguese]

71. Pressinotti, M.M.N, Fernandes da Silva, P.C.; Marchiori-Faria, D.G., & Mendes, R.M. (2009). The experience of the Geological Institute in hazard maping and risk analysis in urban environments. Proceedings of the 11th Southeast Brazil Geological Symposia. São Pedro, November 2009. ISSN 2175-697X. P. 156-157. [in Portuguese] Rossini-Penteado, D., Ferreira, C.J., & Giberti, P.P.C. (2007). Quantification of vulnerability and potential damage applied to hazard mapping and risk analysis, 1:10,000,

72. Ubatuba-SP. Proceedings of the 2nd Brazilian Symposium on Natural and Technological Disasters. Santos, December 2007. [in Portuguese]

73. Sabins Jr., F.F. (1987). Remote Sensing. Principles and Interpretation. Freeman, New York. Soares, P.C. & Fiori, A.P. (1976). Systematic procedures for geological interpretation of aerial photographs. Notícia Geomorlógica 16 (32): 71-104. [In Portuguese] Soares, P.C., Mattos, J.T., Barcellos, P.E., Meneses, P.R., & Guerra, S.M.S. (1981). Regional morpho-structural analysis on Radar and Landsat images at the Parana Basin. Proceedings of the 3rd South Brazilian Regional Symposium on Geology, Curitiba. p. 5-23.[In Portuguese]

74. Thompson, A., Hine, P.D., Poole J.S. & Greig, J.R. (1998a) Environmental Geology in LandUse Planning: Advice for planners and developers.

Published by Symonds Travers Morgan (East Grinstead, UK) on behalf of Department of Environment, Transport and the Regions, Great Britain.

75.   Thompson, A., Hine, P.D., Poole J.S. & Greig, J.R. (1998b) Environmental Geology in Land Use Planning: A guide to good practice. Published by Symonds Travers Morgan (East Grinstead, UK) on behalf of Department of Environment, Transport and the Regions, Great Britain.

76.   Thompson, A., Hine, P.D., Greig, J.R. & Poole J.S. (1998c) Environmental Geology in Land Use Planning: Emerging issues. Published by Symonds Travers Morgan (East Grinstead, UK) on behalf of Department of Environment, Transport and the Regions, Great Britain.

77.   Thornton, S.F., Lerner, D.N., & Banwart, S.A. (2001) Assessing the natural attenuation of organic contaminants in aquifers using plume-scale electron and carbon balances: model development with analysis of uncertainty and parameter sensitivity. Journal of Contaminant Hydrology, 53(3-4): 199-232. ISSN 0169-7722.

78.   Tominaga, L.K., Rossini-Penteado, D., Ferreira, C.J., Vedovello, R. & Armani, G. (2008). Landsliding hazard assessment through the analysis of multiple geo-environmental factors. Proceedings of the 12th Brazilian Congress on Environmental and Engineering Geology. ABGE, ISBN 978-85-7270-052-8. Porto de Galinhas, November 2008. 15p. [In Portuguese]

79.   Tominaga, L.K. (2009a). Natural Disasters: Why they happen ? In: Disasters: Knowing to prevent. Tominaga, L.K., Santoro, J., & Amaral, R (eds). Pp. 147-160, Geological Institute, ISBN 978-85-87235-09-1, São Paulo. [In Portuguese]

80.   Tominaga, L.K. (2009b). Risk Analysis and Mapping. In: Disasters: Knowing to prevent. Tominaga, L.K., Santoro, J., & Amaral, R (eds). Pp. 13-23, Geological Institute, ISBN 978-85-87235-09-1, São Paulo. [In Portuguese]

81.   TRRL - Transport & Road Research Laboratory. (1978). Terrain evaluation for highway engineering and transport planning: a technique with particular value for developing countries. TRRL Supplementary Report No. 448, 21p.

82.   UN-ISDR – United Nations – International Strategy for Disaster Reduction. (2004). Living with Risk: A Global Review of Disaster Reduction Initiatives. In: Inter-Agency Secretariat International Strategy for Disaster Reduction. Geneve, 152p. Accessed in Aug, 2009. Available at http://www.unisddr.org.

83.   Varnes, D.J. (1984). Landslide Hazard Zonation: Review of Principles and Practice. Unesco Press, Paris, 56p.

84. Vedovello, R. (2000). Geotechnical Zoning for Environmental Management through Basic Physiographic Units. (unpublished PhD Thesis, Sao Paulo State University at Rio Claro). [In Portuguese]

85. Vedovello, R. (1993). Geotechnical Zoning based on Remote Sensing Techniques for Urban and Regional Planning. São José dos Campos, 186p. (unpublished M.Sc. Dissertation, National Institute for Space Research) [In Portuguese]

86. Vedovello, R. & Mattos, J.T. (1998). The use of basic physiographic units for definition of geotechnical units: a remote sensing approach. Proceedings of the 3rd Brazilian Symposium on Geotechnical Cartography. São Paulo, November 1998, 11p. [In Portuguese]

87. Vedovello, R., Brollo, M.J., & Fernandes da Silva, P.C. (1998). Assessment of erosion as a conditioning factor on the selection of sites for industrial waste disposal: an approach based on regional physiographic compartmentalisation obtained from satellite imagery. Proceedings of the 3rd Brazilian Symposium on Erosion Control. Presidente Prudente, March 1998. 9p. [In Portuguese]

88. Vedovello, R., Tominaga, L.K., Moura-Fujimoto, N.S.V., Fernandes da Silva, P.C., Holl, M.C., & Maffra, C.Q.T. (1997). A mass movement hazard analysis approach for public authority responses. In: Proceedings of the 1st Latin American Forum on Applied Geography. Curitiba, September 1997. 9p. [In Portuguese]

89. Verstappen HTh (1977) Remote Sensing in Geomorphology. Elsevier Sci. Publ., Amsterdam. 214p.

90. Vinogradov, B.V., Gerenchuk, K.I., Isachenko, A.G., Raman, K.G., & Teselchuk, Y.N. (1962). Basic principles of landscape mapping, Soviet Geography: Review and Translation, 3 (6), 15-20.

91. Vrba J. & Civita, M. (1994) Assessment of groundwater vulnerability. In: Guidebook on Mapping Groundwater Vulnerability. Vrba, J. & Zaporozec, A. (eds.). 16: 31-48, International Association of Hydrogeologists (IAH).

Chapter 2

MINERAL PROSPECTIVITY MAPPING METHOD INTEGRATING MULTI-SOURCES GEOLOGY SPATIAL DATA SETS AND CASE-BASED REASONING

Binbin He[1], Jianhua Chen[2], Cuihua Chen[3], Yue Liu[3]

[1]School of Resources and Environment, University of Electronic Science and Technology of China, Chengdu, China
[2]College of Geophysics, Chengdu University of Technology, Chengdu, China
[3]College of Geosciences, Chengdu University of Technology, Chengdu, China

ABSTRACT

Extracting and synthesizing information from existing and massive amounts of geology spatial data sets is of great scientific significance and has considerable value in its applications. To make mineral exploration less expensive, more efficient, and more accurate, it is important to move beyond traditional concepts and establish a rapid, efficient, and intelligent method of predicting the existence and location of minerals. This paper describes a case-based reasoning (CBR) method for mineral prospectivity mapping that takes spatial features of geology data into account and offers an intelligent approach. This method include a metallogenic case representation that combines spatial and attribute features, metallogenic case-based storage organization, and a metallogenic case similarity retrieval model. The experiments were performed in the eastern Kunlun Mountains, China using CBR and weights-of-evidence (WOE), respectively. The results show that the prediction accuracy of the CBR is higher than that of the WOE.

INTRODUCTION

Mineral prospectivity analysis and quantitative resource estimation have been recognized as important when integrating multi-source geology spatial data in recent years [1]. The statistical and mathematical approaches developed recently for multi-resources geological spatial data integration include weights-of-evidence (WOE) [2-8], and the logistic regression [9,10]. The fuzzy logic [11, 12], artificial neural networks [13,14] and the Fractal method [15] have been applied in the assessment of mineral resources potential. Although these methods promote the efficiency and effectiveness of mineral resource prospecting, their algorithms are unable to accumulate knowledge, and lack intelligent reasoning. Meanwhile, similar deposit types occur in similar geological conditions and spatial distributions. The metallogenic geological conditions and spatial distribution of discovered and typical deposits can be used to construct a historical case-base for mineral prospectivity analysis. Traditional analysis methods cannot mine the depth of information or make intelligent inferences. In recent years, some researchers have begun applying case-based reasoning (CBR) to the environment, urban planning, and land use.

Lekkas et al. [16] suggested using CBR to predict air pollution levels; Holt and Benwell [17] tried using CBR to classify soil; Ye et al. [18] integrated CBR and GIS for urban planning approval; and Du et al. [19] applied CBR for land use change prediction. CBR is a branch of artificial intelligence that began in the research of Schank and Abelson [20]. CBR does not require a precise domain model, and it solves new cases by using historical knowledge. Its application is based on two assumptions about the objective world: 1) similar problems have similar solutions; and 2) similar problems may recur. CBR uses the principle of similarity to find strategies for new cases; it also offers a method that resembles the human problem-solving approach of extracting and storing expertise. From a methodological point of view, CBR proposes a comprehensive, problem-oriented approach to analysis that is more adaptable than rule-based and model-based reasoning. CBR is particularly suitable for areas in which it is difficult to summarize, abstract, and express expertise; this makes CBR useful for solving ambiguous problems. CBR can do quantitative analysis and prediction without a careful mechanism study, and it has advantages in the simplification of knowledge acquisition, the improvement of efficiency and quality, and the accumulation of knowledge. Additionally, CBR and the identification process are highly automated and reusable. CBR is an effective method in cases in which prior knowledge is lacking or for constructing complex issues in quantitative models.

In this paper, a method for mineral prospectivity mapping was proposed integrating multi-sources geology spatial data sets and case-based reasoning, including a metallogenic case representation model that combines spatial and attribute features, the metallogenic case-feature weights-determining model, metallogenic case-based storage organisation, and a metallogenic case-similarity retrieval model. The experiment was performed in the eastern Kunlun Mountains, China to predict the existence of potential iron deposits using case-based reasoning and weights-of-evidence, respectively.

METHODOLOGY

The mineral prospectivity mapping method using casebased reasoning include three main components: a metallogenic case representation model, metallogenic case storage, and a metallogenic case retrieval model. Figure 1 describes the flow of mineral prospectivity mapping method using CBR.

Metallogenic Case Representation Model

Generally, a case in a traditional CBR model is composed of both attribute and goal features. Because of the spatial distribution and regional laws of geological entities, the case representation is different from a traditional one. The features of a metallogenic case include both spatial and attribute features, which are selected or extracted from metallogenic entities.

During the construction of a metallogenic case representation model, each grid of a certain size is taken as a representative object. First, typical feature attributes related to ore control that are contained in vector grids of existing mineral points are extracted. Then, the corresponding names of mines in vector grids and relevant result values are determined. The extracted-features attribute, the corresponding names, and the relevant results are all described by the rules of case expression. To extract spatial features, the orientation relations, the metric relations, and the topology relations related to ore control in each vector grid are extracted, and spatial relations are transformed to attribute mode. Therefore, a metallogenic case consists of general attributes and spatial-relation property items. The basic expression is as follows:

$$C = \left(A_{a1}, A_{a2}, \cdots, A_{ak}, A_{s1}, A_{s2}, \cdots A_{sm}, \text{Result} \right) \tag{1}$$

where A_{ai} is the general feature property item, A_{sj} is the spatial-relation feature property item, and Result represents the result of the case. To solve a new case, existing cases can be extracted by spatial relations under certain rules (e.g., spatial coding). After that, candidates for a historical case set are obtained.

Metallogenic Case Storage

After a typical metallogenic case is constructed, it is stored in a spatial database in database tables or into document systems in a text file. The stored cases are then indexed to improve the efficiency of the metallogenic case-similarity retrieval model.

Metallogenic Case Retrieval Model

Because a metallogenic case has spatial features, it is different than a traditional CBR model. First, during the construction of a metallogenic case retrieval model, all vector grids are set as unsolved cases under the metallogenic case representation model. In other words, each case describes typical attribute and spatial features, and the results description (i.e., the case-determining attribute) is set to blank. Second, a similarity-measure threshold is set, and each unsolved case is retrieved for similarity. After a similar case is found, its result is assigned to the unsolved case according to the threshold and the strategy given. If the case obtained is unsatisfactory, it can be modified by expertise. Its result can then be assigned to the unsolved case. The retrieval unsolved cases in all vector grids are then completed. Third, the typical cases obtained or modified can be stored into the case base for expansion and update.

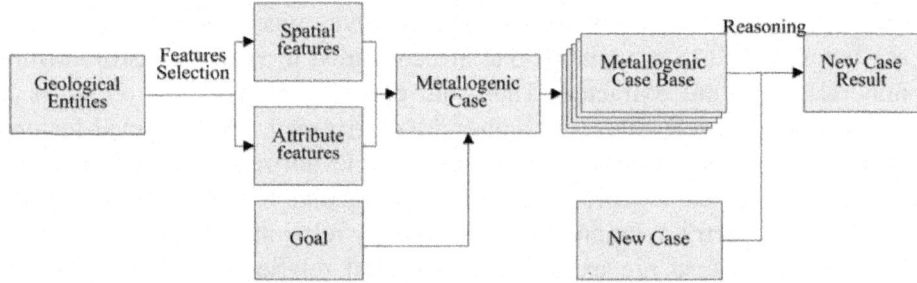

Figure 1: The flow of mineral prospectivity mapping method using CBR.

After a metallogenic case base is constructed, the metallogenic case retrieval model (Figure 2) can use it to compare existing metallogenic cases with new ones. The similarity measurement formulas for existing and new cases are as follows:

$$S_{\%} = \left(100 * \left(1 - \mathrm{sqrt}\left(\frac{\mathrm{distance}}{\mathrm{sum(weights)}}\right)\right)\right)$$
$$* \left(\frac{\mathrm{searchedWeightsSum}}{\mathrm{totalWeightsSum}}\right) \tag{2}$$

$$\mathrm{distance} = \mathrm{weight}_1 * \mathrm{dist}_1^2 + \mathrm{weight}_2 * \mathrm{dist}_2^2$$
$$+ \cdots + \mathrm{weight}_n * \mathrm{dist}_n^2 \tag{3}$$

$$\mathrm{dist} =$$
$$\min\left(1, \frac{\mathrm{diff}\left(\mathrm{newCaseValue,\ caseValue}\right)}{\left(\mathrm{maxValue} - \mathrm{minValue}\right) * \mathrm{infinityCons\,tan}\,t}\right) \tag{4}$$

where "$S_\%$" is similarity ranged between 0% and 100%; "distance" is the weighted sum of the squares of "$dist_i$" ranged between 0 and 1; "searchedWeightsSum" is the sum of the weights, with the new case feature and the actual case feature both being non-empty; "totalWeightsSum" is the sum of the weights of all case features; "$dist_i$" is the distance between the new case feature and the actual case feature, in which the value is the smaller of either 1 or the Euclid distance between the new case feature and the actual case feature; "newCaseValue" is the new case feature value; "caseValue" is the actual case feature value; "maxValue" and "minValue" are the case corresponding feature's maximum and minimum values; and "infinityConstant" is a large constant.

To measure similarity, each new case is compared with all cases in the case base. The return value is based on the selection strategies of the maximum, threshold, or K nearest neighbors. If the value is unsatisfactory, it can be modified by the return value and relevant expertise. The typical cases obtained and the cases modified can be stored into the case base for expansion.

EXPERIMENTS

To verify the effectiveness of the proposed method, the experiments of mineral potential prediction for iron deposits were performed in the eastern Kunlun Mountains, using the metallogenic CBR model and the weights-of-evidence model, respectively. All of the data sets used in this paper were derived from our established multisource geology spatial database, which contains geological, geophysical, geochemical, and remote-sensing data. The metallogenic CBR model was implemented with $C^\#$ based on ArcEngine GIS components. The weights-of-evidence model was performed with Arc-SDM [21].

Geological Setting of Study Area

The eastern Kunlun Mountains are within Qinghai Province, China, and are shown as an insertion from left to right to the provincial map (Figure 3). The Mountains are within latitudes 34°57′ and 37°56′N, and longitudes 90°31′ and 100°04′E. Of the study area, the eastern Kunlun orogenic belt is attached to the southern margin of the Qaidam Basin.

The area consists of three major deep crustal-scale faults that divide the area roughly from north to south into subtectonic belts (Figure 4). Kunbei ("Kun" is short for Kunlun. "bei" means north in Chinese) belt is in the north. It belongs to the Kunbei Caledonian back-arc basin situated mainly in the northwestern part of the Kunlun Mountains. The belt is made of early Palaeozoic folding belts dominated by the Ordovician marine sediments

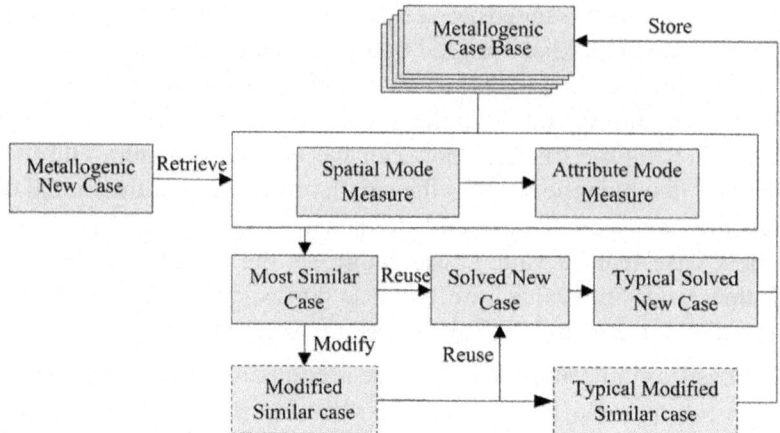

Figure 2: Metallogenic case similarity retrieval model.

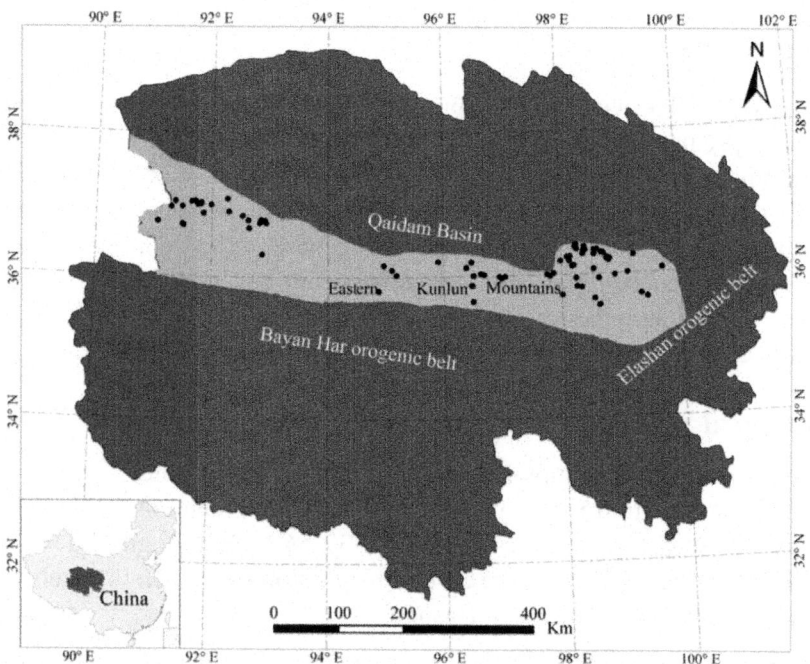

Figure 3: Eastern Kunlun Mountains within Qinghai Province, China.

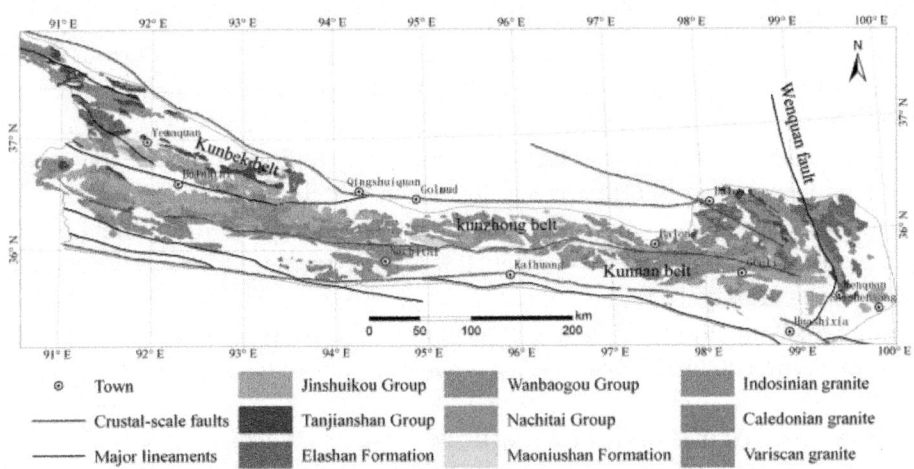

Figure 4: A simplified geological map of the eastern Kunlun Mountains, China.

Note: Showing major lithologic units, stratigraphic units, and crustal-scale faults. The east Kunlun orogenic belt is subdivided into the Kunbei belt, and the Kunnan belt. and low-grade metamorphic rock. Kunzhong ("zhong" is middle in Chinese) belt is the basement of an uplift belt and a granitic belt. It is made predominantly of the middle to late Proterozoic metamorphic sequences, and Palaeozoic and Mesozoic granitic rock. The Devonian continental sandstones, conglomerate, and volcanic rock, and Carboniferous marine limestone and sedimentary rock lie over the metamorphic and plutonic basement. The composition of Kunnan ("nan" means south in Chinese) belt is geologically similar to that of the Kunzhong belt, but it consists of numerous Triassic successions. As of today, there are 81 known sites of iron formation within the area. Their locations are shown as black dots in Figure 3.

Within the study area, regions exposed mainly by lithologic and stratigraphic units are displayed inFigure 4. The Jinshuikou Group is the oldest crystalline basement that comprises gneiss, amphibolite rock, migmatite, and marble. It belongs to a suite of middle-to-high grade of metamorphic rock [22]. The Tanjianshan Group of the Ordovician-Cambrian period is composed of intermediate-mafic volcanic rock, and phyllite crystalline limestone and sandstone. The Elashan Formation of the Triassic time consists basically of volcanic rock that is intermediate-acid. The rock is with sandstone intercalation. The Wanbaogou Group of the New Mesoproterozoic period is subdivided into an upper unit and a lower one, comprising mainly carbonate rock and intermediate-mafic volcanic rock. Both types of rock belong to the pre-Cambrian folding basement with a low-grade of metamorphism [23].

The volcanic rock and carbonaceous slate of the Wanbaogou Group serve as important ore beds of precious metal (e.g., gold) and non-ferrous metals (copper, cobalt, and nickel) in the Kunnan belt [24]. The Nachitai Group of the Ordovician period consists largely of schist, mafic volcanic rock, chert, and crystalline limestone. The Maoniushan Formation of the Devonian time is composed of an intermediate-acid volcanic rock underlain by clastic rock. The Variscan-Indosinian granite is closely associated with the metalliferous mineralization in the region when the granite occurs extensively, diversely and permanently [25]. Known iron mineralization occurred mainly in the Yemaquan metallogenic belt located in the western part of the Mountains, whereas the Dulan-Elashan metallogenic belt lies in the east.

Data Preprocessing and Metallogenic Case Construction

The best ore-controlling variable and threshold were determined using proximity analysis of the weights-ofevidence model. On the basis of a correlation analysis among evidence variables, the authors selected vector ore-controlling data of stratum, unconformity, fault, regional geochemical data, remote-sensing mineralization information, Bouguer gravity data, aeromagnetic dataand mineral occurrence for this experiment. Before constructing the specific metallogenic case, the region was partitioned into 96,576 grids, each one being 1 km by 1 km. All of the evidence-variable data were spatially joined to a grid polygon, and each grid had corresponding feature-attribute values. The unconformity, fault, and mineral occurrences were buffered by with distances of 3000 m, 300 m, and 1000 m, respectively. To extract the spatial features of the metallogenic case, the fault's direction (orientation relationship), the shortest distance between mineral occurrence and faults (metric relationship), the disjoint relationship between mineral occurrence and faults, and the unconformities (topology relationship) were computed and extracted in each grid polygon. The spatial relationships were then transformed into attributes and stored in the attribute tables of each grid polygon. This process paved the way for metallogenic case retrieval. In this way, the metallogenic representation model combined with spatial and attribute features was constructed. Each grid polygon became a case representation object. By analyzing each grid layer's attribute table in tandem with ore-controlling factors, the authors established the metallogenic case's attribute features by using lithological characters, chronostratigraphy, unconformity, fault, regional chemical anomaly, remote-sensing mineralisation anomaly, Bouguer gravity anomaly, and aeromagnetic anomaly. In this way, specific genetic types became object attributes. The metallogenic case representation model in this research is as follows:

C = (unconformity, regional geochemical anomaly, Bouguer gravity anomaly, aeromagnetic anomaly, chronostratigraphy, lithological characters, remote-sensing mineralisation anomaly, fault characters, fault directions, short distance to fault, distance to unconformity, disjoint fault, disjoint unconformity, genetic type).

Prior to analysis, the attribute and spatial features of the above case are set corresponding weights, which are determined and assigned based on the Analytic Hierarchy Process (AHP) [26]. On the basis of expert knowledge, the importance of AHP case features is as follows: regional geochemical anomaly > fault directions > short distance to fault = disjoint fault > fault characters = remote-sensing mineralisation anomaly > chronostratigraphy = lithological characters > distance to unconformity = disjoint unconformity > unconformity characters = Bouguer gravity anomaly = aeromagnetic anomaly. Table 1 shows the comparison matrix of metallogenic CBR features by AHP. The matrix is equalised and simplified to seven features. After calculating, uniformity has been passed and each feature weight determined; identical, important features have the same weights (Table 2).

To grid the polygonal layers that are overlapped by attribute and spatial features, the authors analysed each grid

Table 1: The CBR features comparison matrix by AHP

	B1	B2	B3	B4	B5	B6	B7	B8	B9	B10	B11	B12	B13
B1	1	1/7	1	1	1/3	1/3	1/4	1/4	1/6	1/5	½	1/5	1/2
B2		1	7	7	7/3	7/3	7/4	7/4	7/6	7/5	7/2	7/5	7/2
B3			1	1	1/3	1/3	1/4	1/4	1/6	1/5	½	1/5	1/2
B4				1	1/3	1/3	1/4	1/4	1/6	1/5	½	1/5	1/2
B5					1	1	3/4	3/4	3/6	3/5	3/2	3/5	3/2
B6						1	3/4	3/4	3/6	3/5	3/2	3/5	3/2
B7							1	1	4/6	4/5	4/2	4/5	4/2
B8								1	4/6	4/5	4/2	4/5	4/2
B9									1	6/5	6/2	6/5	6/2
B10										1	5/2	1	5/2
B11											1	2/5	1
B12												1	5/2
B13													1

B1: unconformity; B2: regional geochemical anomaly; B3: Bouguer gravity anomaly; B4: aeromagnetic anomaly; B5: chronostratigraphy; B6: lithological characters; B7: remote-sensing mineralisation anomaly; B8: fault characters; B9: fault directions; B10: short distance to fault; B11: distance to unconformity; B12:disjoint fault; B13:disjoint unconformity.

Table 2: The CBR equivalent property features comparison matrix and weights determined by AHP

	B1	B2	B3	B4	B5	B6	B7	Weights
B1	1	7	7/3	7/4	7/6	7/5	7/2	0.250
B2		1	1/3	¼	1/6	1/5	1/2	0.036
B3			1	¾	3/6	3/5	3/2	0.107
B4				1	4/6	4/5	4/2	0.143
B5					1	6/5	6/2	0.214
B6						1	5/2	0.179
B7							1	0.071

B1: regional geochemical anomaly; B2: Bouguer anomaly; B3: chronostratigraphy; B4: remote-sensing mineralization anomaly; B5: fault directions; B6: short distance to fault; B7: distance to unconformity.

layer's attribute table, selected all the records in which the field showing the genetic type of mineral occurrence was non-empty, and exported those records for further analysis. The final records were stored in a text file in which all attribute values were separated by tabs. The corresponding genetic type case base was then constructed. The attribute tables of relevant grid polygon layers were exported and stored in a text file, and each unsolved case set was constructed (each grid represents an unsolved case object). Each grid in the polygon layers corresponds to an unsolved metallogenic case. After a similarity measurement, each grid was assigned a genetic type, and the similarities were assigned values between 0 and 100%. In this way, the classification strategy automatically outlined a regional metallogenic prediction map showing high, medium, and low potentials.

Mineral Potential Prediction Results and Analysis

Based on the data-processing and metallogenic CBR model described above, an experiment regarding mineral potential prediction for iron deposits was performed in the eastern Kunlun Mountains, China. Figure 5 reports, respectively, the curves representing the relationships between 1) posterior probability based on the WOE and cumulative mineral occurrence; and 2) posterior probability and cumulative areas. Table 3 and Figure 6 show the favorable metallogenic potential regions (i.e., areas of high and medium potential) extracted using weights-ofevidence model. Highand medium-potential areas occupy 21% of the study area and contain 62 points of 81 known deposit points (i.e., 77% of known deposit points). High-potential areas occupy 11% of the total area and include 45 known deposit points (i.e., 56% of known deposit points). Medium-potential areas occupy 10% of the total area and include 17 known deposit points (i.e., 21% of known deposit points).

Table 3 and Figure 7 present the potential prediction results for iron deposits using the proposed metallogenic CBR method. Favorable metallogenic regions (i.e., highand medium-potential areas) account for 21% of the

Table 3: The contrast of prediction results using CBR and WOE

Method	high potential (number, percent)	medium-potential (number, percent)	low-potential (number, percent)
WOE	45, 56%	17, 21%	19, 23%
CBR	68, 84%	5, 6%	8, 10%

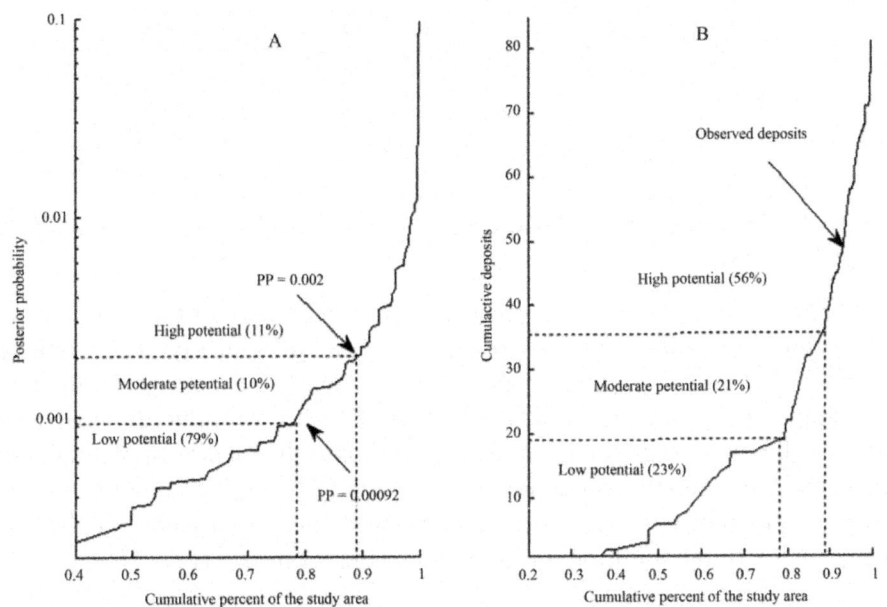

Figure 5: Variation of cumulative area with sum of weights and cumulative deposits using WOE.

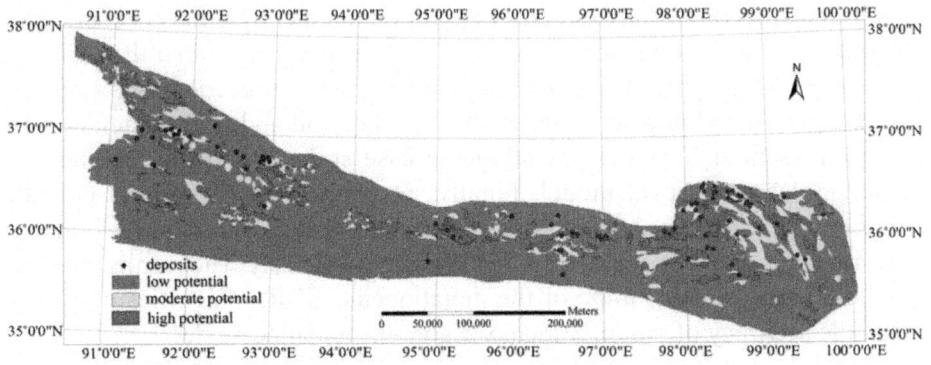

Figure 6: Potential prediction map for iron deposits using WOE in eastern Kunlun Moutains, China.

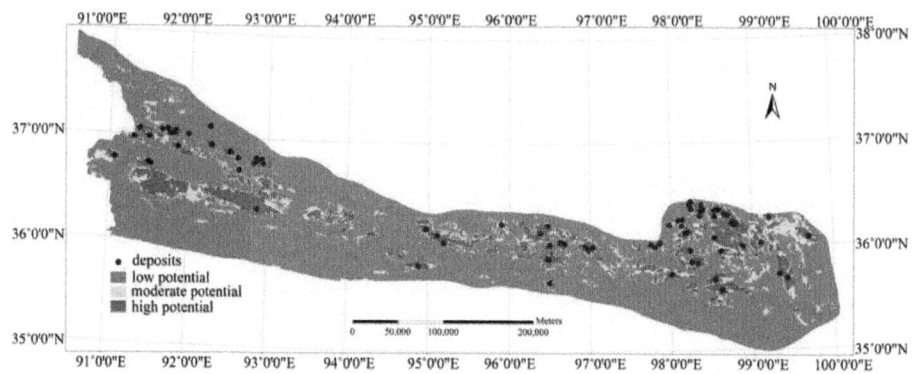

Figure 7: Potential prediction map for iron deposits using CBR in eastern Kunlun Mountains, China.

study area, with high-potential areas accounting for 10% of the total area and the medium-potential areas accounting for 11% of the total area. The prediction results show that known mineral occurrence is highly consistent with the high-potential areas, as analysis predicts that 68 of 81 known mineral occurrences fall into the high-potential areas (84%), 5 fall into the medium-potential areas (6%), and 8 fall into the low-potential areas (10%). Overall prediction accuracy (highand medium-potential areas account for 90%) is significantly higher than the accuracy of the traditional weights-of-evidence model (i.e., 77%).

CONCLUSION

The metallogenic CBR method for regional mineral prospectivity mapping is a new and intelligent prediction method. It makes full use of multisource massive geology spatial data. It also surpasses traditional mineralprediction approaches to improve the intelligence, efficiency, and accuracy of mineral prediction. This paper takes spatial features of geology data into account and proposes an integral metallogenic CBR method, which includes the metallogenic case representation model, metallogenic case storage, and the metallogenic case similarity retrieval model. Finally, an application of mineral potential prediction for iron deposits was performed in the eastern Kunlun Mountains, China, using a metallogenic CBR and WOE, respectively. The results indicated that the prediction accuracy of the metallogenic CBR is significantly higher than the accuracy of the traditional weights-of-evidence model.

ACKNOWLEDGEMENTS

This study was supported by the National Natural Science Foundation of China (Grant No. 41171302 & 40701146) and the National High-Tech R&D Program of China (Grant No. 2007AA12Z227). We express our gratitude to Zhuang Yongcheng from the Qinghai Institute of Geology Survey for his directions.

REFERENCES

1. K. Porwal and O. P. Kreuzer, "Introduction to the Special Issue: Mineral Prospectivity Analysis and Quantitative Resource Estimation," Ore Geology Reviews, Vol. 38, No. 3, 2010, pp. 121-127. doi:10.1016/j. oregeorev.2010.06.002

2. G. F. Bonham-Carter, F. P. Agterberg and D. F. Wright, "Weights of Evidence Modeling: A New Approach to Mapping Mineral Potential," In: F. P. Agterberg, et al., Eds., Statistical Applications in the Earth Sciences, Canadian Government Publishing Centre, Ottawa, 1989, pp. 171-183.

3. F. P. Agterberg, "Combining Indicator Patterns in Weights of Evidence Modeling for Resource Evaluation," Natural Resources Research, Vol. 1, No. 1, 1992, pp. 39-50.doi:10.1007/BF01782111

4. J. R. Harris, L. Wilkinson and E. C. Grunsky, "Effective Use and Interpretation of Lithogeochemical Data in Regional Mineral Exploration Programs: Application of Geographic Information Systems (GIS) Technology," Ore Geology Reviews, Vol. 16, No. 3-4, 2000, pp. 107-143. doi:10.1016/S0169-1368(99)00027-X

5. E. J. M. Carranza, "Weights of Evidence Model of Mineral Potential: A Case Study Using Small Number of Prospects, Abra, Philippines," Natural Resources Research, Vol. 13, No. 3, 2004, pp. 173-187. doi:10.1023/ B:NARR.0000046919.87758.f5

6. Daneshfar, A. Desrochers and P. Budkewitsch, "Mineral-Potential Mapping for MVT Deposits with Limited Data Sets Using Landsat Data and Geological Evidence in the Borden Basin, Northern Baffin Island, Nunavut, Canada," Natural Resources Research, Vol. 15, No. 3, 2006, pp. 129-149. doi:10.1007/s11053-006-9020-7

7. Porwal, I. González-Álvarez, V. Markwitz, T. C. McCuaig and A. Mamuse, "Weights-of-Evidence and Logistic Regression Modeling of Magmatic Nickel Sulfide Prospectivity in the Yilgarn Craton, Western Australia," Ore Geology Reviews, Vol. 38, No. 3, 2010, pp. 184-196. doi:10.1016/j. oregeorev.2010.04.002

8. B. He, C. H. Chen and Y. Liu, "Gold Resources Potential Assessment in Eastern Kunlun Mountains of China Combining Weights-of-Evidence Model with GIS Spatial Analysis Technique," Chinese Geographical Science, Vol. 20, No. 5, 2010, pp. 461-470.doi:10.1007/s11769-010-0420-6

9.   F. P. Agterberg, G. F. Bonham-Carter, Q. M. Cheng and D. F. Wright, "Weights of Evidence Model and Weighted Logistic Regression in Mineral Potential Mapping," In: J. C. Davis, et al., Eds., Computers in Geology, Oxford University Press, New York, 1993, pp. 13-32.

10. E. J. M. Carranza and M. Hale, "Logistic Regression for Geologically Constrained Mapping of Gold Potential, Baguio District, Philippines," Exploration and Mining Geology, Vol. 10, No. 3, 2001, pp. 165-175. doi:10.2113/0100165

11. F. Aminzadeh, "Applications of Fuzzy Expert Systems in Integrated Oil Exploration," Computers & Electrical Engineering, Vol. 20, No. 2, 1994, pp. 89-97. doi:10.1016/0045-7906(94)90023-X

12. X. Luo and R. Dimitrakopoulos, "Data-Driven Fuzzy Analysis in Quantitative Mineral Resource Assessment," Computers & Geosciences, Vol. 29, No. 1, 2003, pp. 3-13.doi:10.1016/S0098-3004(02)00078-X

13. K. Koike, S. Matsuda, T. Suzuki and M. Ohmi, "Neural Network-Based Estimation of Principal Metal Contents in the Hokuroku District, Northern Japan, for Exploring Kuroko-Type Deposits," Natural Resources Research, Vol. 11, No. 2, 2002, pp. 135-156.doi:10.1023/A:1015520204066

14. J. P. Rigol-Sanchez, M. Chica-Olmo and F. Abarca-Hernandez, "Artificial Neural Networks as a Tool for Mineral Potential Mapping with GIS," International Journal of Remote Sensing, Vol. 24, No. 5, 2003, pp. 1151-1156. doi:10.1080/0143116021000031791

15. P. Gumiel, D. J. Sanderson, M. Arias, S. Roberts and A. Martín-Izard, "Analysis of the Fractal Clustering of Ore Deposits in the Spanish Iberian Pyrite Belt," Ore Geology Reviews, Vol. 38, No. 4, 2010, pp. 307-318. doi:10.1016/j.oregeorev.2010.08.001

16. G. P. Lekkas, N. M. Avouris and L. G. Viras, "CaseBased Reasoning in Environmental Monitoring Applications," Applied Artificial Intelligence, Vol. 8, No. 3, 1994, pp. 359-376.doi:10.1080/08839519408945448

17. Holt and G. L. Benwell, "Applying Case-Based Reasoning Techniques in GIS," International Journal of Geographical Information Science, Vol. 13, No. 1, 1999, pp. 9-25.doi:10.1080/136588199241436

18. J. A. Ye and X. Shi, "Integrating Case-Based Reasoning and GIS for Handling Planning Applications," Journal of Urban Planning Department, No. 3, 2001, pp. 34-40.

19. Y. Y. Du, W. Wen, F. Cao and J. Min, "A Case-Based Reasoning Approach for Land Use Change Prediction," Expert Systems with Applications, Vol. 37, 2010, pp. 5745-5750.doi:10.1016/j.eswa.2010.02.035

20. R. C. Schank and P. A. Robert, "Scripts, Plans, Goals, and Understanding: An Inquiry into Human Knowledge Structures," Lawrence Erlbaum, Hillsdale, 1977.

21. L. D. Kemp, G. F. Bonham-Carter and G. L. Raines, "ArcWofE: ArcView Extension for Weights of Evidence Mapping [EB/OL]," 1999. http://gis. nrcan.gc.ca/software/arcview/wofe

22. G. C. Wang, Q. H. Wang, P. Jian and Y. H. Zhu, "Zircon SHRIMP Ages of Precambrian Metamorphic Basement Rocks and Their Tectonic Significance in the Eastern Kunlun Mountains, Qinghai province, China," Earth Science Frontiers, Vol. 11, No. 4, 2004, pp. 481-490 .

23. Y. S. Pan, W. M. Zhou and R. H. Xu, "The Early Palaozoic Geological Features and Evolutions of the Kunlun Mountain," Science in China (Series D), Vol. 26, No. 4, 1996, pp. 302-307.

24. X. D. Guo, Y. J. Zhang, G. G. Liu, A. J. Pan and F. Zhang, "Metallogenic Regularities and Prospecting Direction of Gold and Copper in Eastern Kunlun," Gold Geology, Vol. 10, No. 4, 2004, pp. 16-22.

25. Z. Z. Qian, Z. G. Hu, J. Q. Liu and H. M. Li, "Active Continental Margin and Regional Metallogenesis of the Palaeo-Tethys in the East Kunlun Mountains," Geotectonica et Metallogenia, Vol. 24, No. 2, 2000, pp. 134-139.

26. T. L. Saaty and L. G. Vargas, "Uncertainty and Rank Order in the Analytic Hierarchy Process," European Journal of Operational Research, Vol. 32, No. 1, 1987, pp. 107- 117.doi:10.1016/0377-2217(87)90275-X

Chapter 3

MAPPING SPATIAL DATA ON THE WEB USING FREE AND OPEN-SOURCE TOOLS: A PROTOTYPE IMPLEMENTATION

Sunil Pratap Singh, Preetvanti Singh

Department of Physics and Computer Science, Dayalbagh Educational Institute, Agra, India

ABSTRACT

There is a growing need for web-based geographic information systems for easy and fast dissemination, sharing, displaying and processing of spatial information. The tremendous growth in the use of web and open-source geospatial resources has sparked development of web-based spatial applications to address multidisciplinary issues with spatial dimensions. This paper presents the integration of open-source geospatial tools and web technology to visualize and interact with spatial data using web browser. The goal of this paper is to implement a prototype system for web-based mapping by providing step-by-step instructions in order to encourage the eager developers and interested readers to publish their maps on the web with no prior technical experience in map servers. The implementation of mapping prototype shows the utilization of open-source geospatial tools which results in a rapid implementation with minimal or no software input cost.

INTRODUCTION

The geographic information systems (GIS) were initially adopted by users in the government sectors, but are now rapidly growing and have profound implications in a large number of application domains including planning [1], marketing [2], retail [3], transportation [4], traffic [5], tourism [6], natural resources [7], real estate [8], agriculture [9] and healthcare [10]. These systems can be valuable planning tools to assist in the development and administration of scientific programs based on geographically identified needs [11]. Linking location to information is a process that applies to many aspects of business such as choosing a site, targeting a market, planning a distribution network or delivery route, drawing up sales territory and allocating resources [12]. The companies, communities and individuals are realizing that they must have geographical information to understand their world.

The tremendous growth in popularity of the internet and the growing public interests in accessing online geo-spatial information have given a move to develop web mapping applications. The web is playing a huge role in spatial application development mainly because of advantages such as platform independency, reduction in distribution costs and maintenance problems, ease of use and widespread access. The most important power of web-based spatial application is the capability to publish and share geo-spatial information on the web which jointly allows information to be exchanged in a rapid and efficient manner thereby helping individuals make important decision quicker. The geo-spatial information not only includes maps or locations of landmarks, but multiple attribute data, socioeconomic data, aerial photographs, satellite images, etc., which may have static or dynamic characteristics [13].

This paper is aimed to provide a method for building a new web-based spatial visualization prototype by using a combination of open-source geospatial packages and Microsoft .NET platform. The following sections specifically document the development and implementation of a prototype system by adopting web-based client-server computing technology.

MATERIALS AND METHODS

Open-Source Geospatial Tools

The commercial software are predominantly "closed source" which means the code cannot be accessed or modified and they may, in some cases, come at a relatively high price with ongoing licensing fees to maintain their use. These systems are thus often priced out of the reach of the resource-constrained areas, particularly within developing countries, which lead them into seeking open-source alternatives [14]. The wide spread free of cost availability, freedom to read, redistribution and modification facilities have made the open-source packages as feature rich alternatives to proprietary software packages [15]. These packages can help the implementation of mapping and spatial analysis tools in a large number of private and public firms, especially in those small organizations that cannot afford proprietary software's cost, complexity, training cost and special requirements. The market of open-source geospatial packages is growing [16] and the quality of these packages is improving with the strong and free collaboration between a large numbers of developers from all around the world.

The open-source project offerings cover the spectrum of tools that interact with geospatial data. The interaction occurs on several fronts: creating, converting, manipulating, analysing or visualizing the geospatial data. These are the common categories of tools or the functional components of open-source geospatial tools. Figure 1 represents the organization of these tools around the geospatial data and some of the well-known geospatial tools are listed in Table 1.

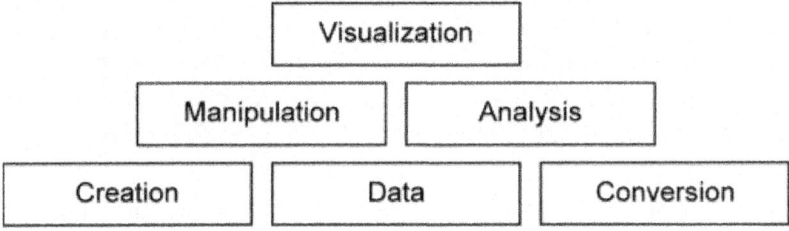

Figure 1: Functional components of open-source tools [17].

Table 1: List of open-source geospatial tools [18]

Category	Open-Source Geospatial Tools
Data Creation	Quantum GIS, OpenEV
Data Conversion	GDAL/OGR, AVCEOO, shp2pgsql (PostGIS)
Data Manipulation	GDAL/OGR, PostGIS, GEOS
Data Analysis	GRASS-GIS, PostGIS
Data Visualization	UMN MapServer, GeoServer, Degree, MapGuide

Microsoft .NET Platform

A range of development technologies is available to develop web based mapping applications. Microsoft's .NET is a revolutionary software programming architecture aimed to create web based applications and offers objectoriented support. The use of object orientation in web applications has increased in recent years since applications are well structured, scalable to new changes and offer better operating performance. The Microsoft .NET Platform provides all the tools and technologies that are needed to build web applications. It exposes a language independent and unified programming model. Its main tools are .NET Framework and Visual Studio .NET (Figure 2).

NET Framework

The .NET Framework is a technology that supports building and running the next generation of web services and applications. It is a managed execution environment that provides a variety of services (memory management, common type system, language interoperability, version compatibility) to its running applications. Its main components include [19]: the common language runtime (CLR), which is the managed execution environment that handles memory allocation, error trapping, and interacting with the operating-system services; the .NET Framework Class Library, which is an extensive collection of programming components and application program interfaces (APIs) for all major areas of application development; ASP.NET, a server side web technology which supports creation of dynamic web applications, websites and web services.

The advantages offered by the .NET Framework include shorter development cycles (code reuse, support for multiple programming languages), easier deployment, fewer data type related bugs due to integral type safety, reduced memory leaks, and, in general more scalable, reliable applications.

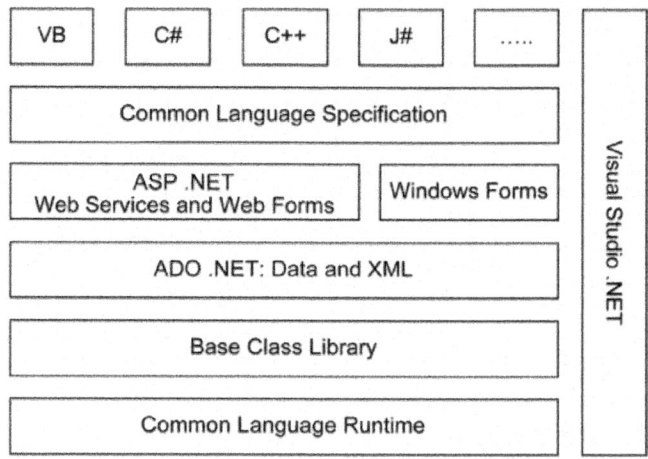

Figure 2: Microsoft .NET platform architecture [19].

Visual Studio .NET

The Microsoft Visual Studio .NET is an integrated development environment (IDE) to develop console and graphical user interface applications along with web applications, web services, websites and windows forms. It supports different programming languages (C#, C++, Visual Basic, J#, etc.) by means of language services, which allow the code editor and debugger to support nearly any programming language, provided a languagespecific service exist. The built-in languages include Visual C++, Visual Basic, Visual C# and Visual J#. It also supports XML/XSTL, HTML/XHTML, JavaScript and CSS. It has comprehensive tools (drag-and-drop form designer, IntelliSense, dynamic help) for rapidly building and integrating applications which increases developer's productivity. The Microsoft Express set of free software applications offer a way around budget limits, as all of the software are free, albeit somewhat restricted. Visual Web Developer 2008 Express is a .NET developer's integrated development environment (IDE) provided for free by Microsoft. It is a lightweight version of Microsoft Visual Studio product line. Figure 3 shows the starting screen of Visual Web Developer 2008 Express. The idea of Express edition is to provide streamlined, easy-to-use and easy-to-learn IDEs for users other than professional software developers. It allows web development with ASP.NET using drag-and-drop user interface designer, enhanced code editor, support for other web technologies (e.g., CSS, JavaScript, XML) and integrated design time validation.

Prototype Development

Preparation of Geospatial Data

The city of Taj, Agra in India is selected as a pilot area to create the prototype spatial mapping system. The geographic location of the scope lies within 27.11° latitude north and 78.20° longitude east and is covered in 1:50,000 scaled sheets 54E/16–54I/4 of Survey of India.

The data that is stored in spatial information system can be thought of as a digital representation of real world objects. These objects can be abstracted into either discrete objects such as houses or continuous fields such as elevation and are stored as vector or raster data structures respectively. The raster model is a representation of the world as a surface that is divided into a regular grid of cells. The vector model represents space as a series of discrete entity-defined point, line or polygon units which are geographically referenced. The vector data models are more compact and predominant across the diverse range of spatial information systems [6,7,20,21] and therefore we have focused on vector model for mapping the geographic data in web-based environment.

The geographic objects together with their attribute data constitute spatial data (known locations on the earth with statistical and non-location data associated with spatial entity). The collection of spatial data under vector model is organized in a thematic approach that categorizes data in layers. Usually, each data layer represents only one type of spatial entities, i.e., point, line or polygon.

There are many sources for obtaining spatial data including hard copy maps; aerial photographs; remotelysensed imagery and existing digital data files. For the scope of this prototype application, a geo-referenced color raster map is obtained in digital format from the Survey of India (government organization). This raster map is processed using Quantum GIS (an open-source desktop mapping engine) to create separate vector layers for different geometry types. Table 2 lists some of the common layers created by extracting geo-objects from the raster map and are used in the prototype application. These layers are saved in ESRI's (Environmental Research System Institute, Redlands, California) shapefile format.

System Components

Database

The analysis on spatial data can be performed well if the data of geographic objects is stored in relational database. For this reason, all the layers (shapefiles) of the study area are converted to their corresponding tables in a PostgreSQL database. PostgreSQL is one of the most popular object-relational database management system (ORDBMS) on the open-source platform. It is not easy to store the spatial data in a standard RDBMS, thus spatial extensions have been developed and standardized by the OGC (Open Geo Consortium).

PostGIS is an open-source and OGC compliant spatial database extender for PostgreSQL [22]. It adds spatial function and specialty geometry data types to the database. It is an excellent way to bring tabular and spatial data together into a common management environment. The conversion of shapefiles to PostgreSQL database tables is achieved using the shp2pgsql utility included as part of the PostGIS extension. This utility takes a shapefile and outputs a series of SQL statements to create a table in PostgreSQL database. The resulting table contains all the attributes of the shapefile including the

Table 2: Layers extracted from base map (raster map)

Sr. No.	Layer	Geometry Type
1	Roads	Line
2	Administrative Buildings	Polygon
3	Points of Interests	Point
4	Railway Lines	Line
5	Public Services (Amenities)	Point

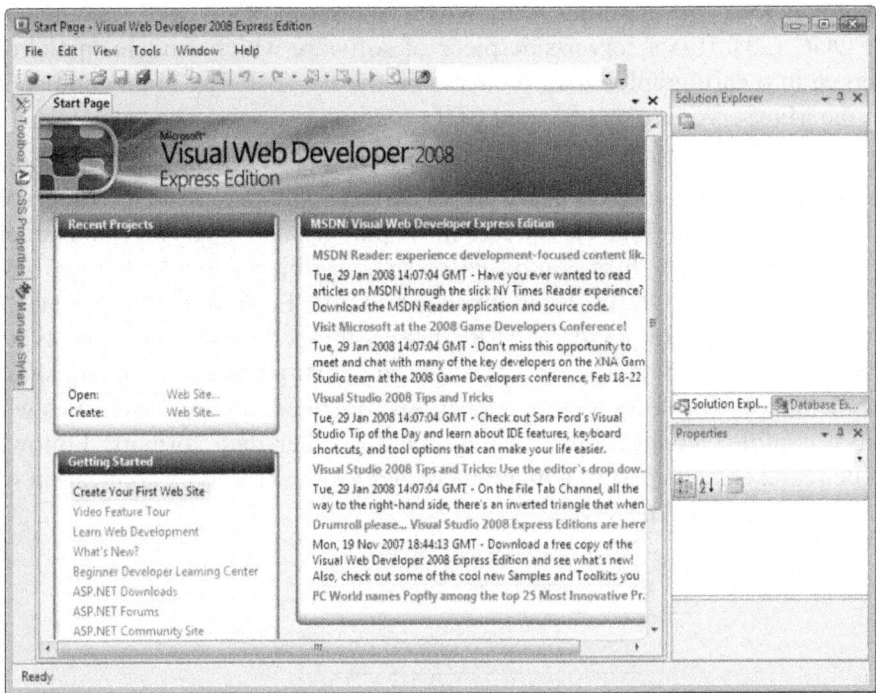

Figure 3: Microsoft Visual Web Developer 2008 Express Edition—start page.

coordinates that define each feature. This process was repeated for each of the vector layers created for the scope of prototype. Figure 4 shows the flow diagram of necessary steps to create the spatial database.

Table 3 represents the open-source geospatial tools used to implement web-based mapping application.

Mapping Server

Internet map server applications easily make spatial data accessible through a web browser interface to end users. The map servers usually serve spatial data based on three OGC standards: WMS (Web Mapping Service) for the display of maps as image; WFS (Web Feature Service) for vector data; and WCS (Web Coverage Service) for raster data. The two best-known open-source map servers are Minnesota MapServer and GeoServer. Both of these are viable alternatives to proprietary map servers.

We have used MapServer for development of current prototype, because it has an excellent performance of functionality and speed on processing large volume dataset. It is considered as the world's leading open-source web mapping tool and complies with WMS, WFS and WCS web specifications of OGC [23]. It is a server-side piece of software which renders spatial data source into cartographic map products on-the-fly [24]. The main components of the MapServer are mapfile and CGI program.

A mapfile defines how data is used and sets the display and query parameters for the map. It needs to set cartographic parameters, cartographic objects, data load- ing, classification, displaying and querying, and graphic elements definition. It is composed of several objects in which each object starts by keyword OBJECT and finishes by END. Each layer object inside the mapfile starts by keyword LAYER, defines the path and connection type to load the specified data, specifies the style and other parameters, and finishes by END. A mapfile allows direct connections for shapefiles and raster files while more complex connections are required for other data formats. Following code represents the part of a mapfile created for the prototype application with connection to PostGIS layer.

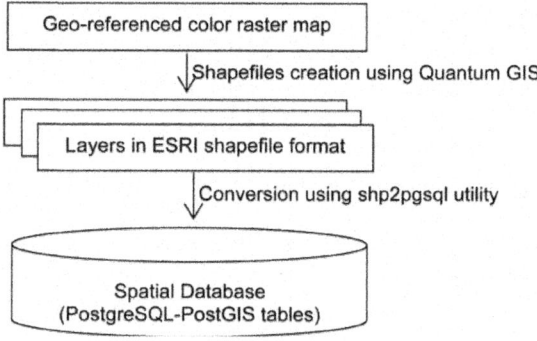

Figure 4: Flow diagram of spatial database development.

Table 3: List of open-source geospatial tools used in study

Category	Open-Source Geospatial Tool
Data Creation	Quantum GIS
Data Conversion	shp2pgsql (PostGIS)
Data Manipulation	PostGIS
Data Analysis	PostGIS
Data Visualization (Internet Mapping)	UMN MapServer

LAYER CONNECTIONTYPE PostGIS CONNECTION "Host = Localhost DbName = dbDemo User = postgresql Password = pgAdmin Port = 5432"

DATA "the_geom FROM POI"

NAME "POI"

TYPE POINT STATUS ON CLASS STYLE COLOR 255 0 0 OUTLINECOLOR 0 0 0 SYMBOL "Circle"

END END END

………..

A CGI program is the real data processing engine: started up by the web server, it reads the map file settings and returns processed spatial data as maps, cartographic objects, variable values and query results.

MapServer can be used in two ways, with the CGI interface or with the MapScript. MapScript provides a scripting interface for MapServer for the development of web based stand-alone applications. It supports popular scripting languages such as PHP, Java, Perl, C# etc. The prototype application in this study is implemented with MapScript API, accessed with C# MapScript under ASP.NET environment. To gain access to PostgreSQL database server, Npgsql is used which is an open-source ASP.NET data provider. The description on how to configure the MapServer and taking reference of precompiled binaries is provided in Appendix I.

Development Environment and User Interface

The user interface design makes the user's interaction simple and efficient in terms of accomplishing user goals. It allows the users to select and input the query criteria in order to view the required data. In the present prototype application, the user interface is made up of ".aspx" files which are populated with the appropriate controls using Visual Web Developer 2008 Express as the .NET development environment. Figure 5 demonstrates the running

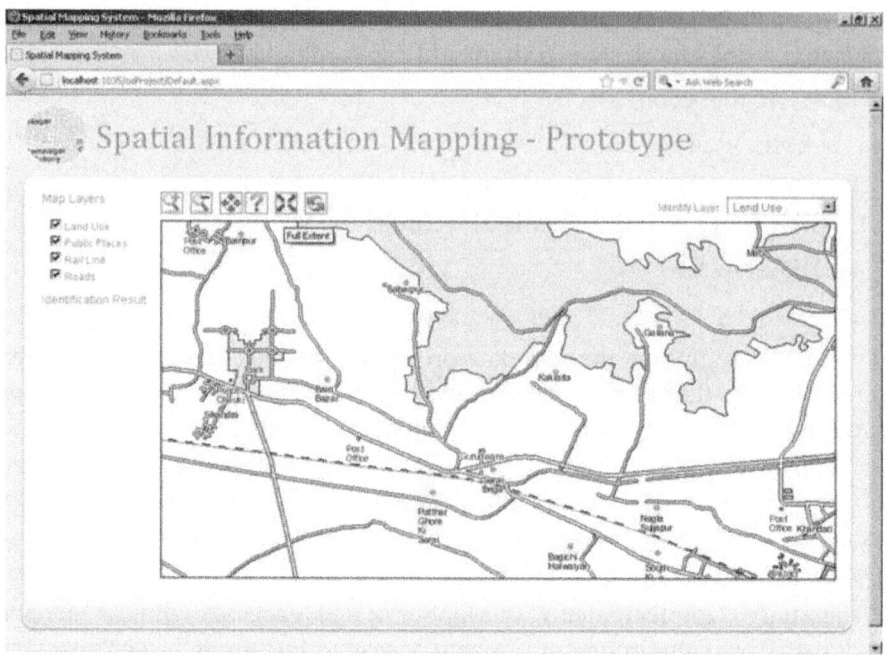

Figure 5: User interface (running view).

view of user interface implemented with ASP.NET by which user interact with the navigation tools such as zoom in, zoom out, zoom to full extent to view the map at different scales and identification of different map features.

System Configuration and Architecture

Hardware Platform Configuration

The server machine used for prototype application development has the following specifications:

CPU: Pentium Dual 3.00 GHz with 2 MB Cache Main Memory: 2048 MB (1024 × 2)

HDD: 80 GB Operating System: Windows XP (SP3)

Web Server: Internet Information Server (IIS) 6.0 DBMS: PostgreSQL with PostGIS

System Architecture

The prototype system adopts client-server architecture where most of the processing takes place on the server side while the client is used for gathering input from the user and displaying map and other associated data based on query parameters. The server-side approach only requires a browser as the spatial mapping and processing; modelling, database and other components will be located on the server. The architecture of the prototype application is shown in the Figure 6, in which the users using any of the popular internet browsers requests geospatial information from the web server.

The web server forwards the request to the ASP.NET application server in which C# Mapscript is called for geospatial data. C# Mapscript forwards the map request to MapServer and the MapServer using the mapfile generates a map based on the geospatial data supplied by the PostGIS. The generated map is sent back to the web server, which forwards it to the web client.

Explanation and Code

The prototype application contains two web forms named Default.aspx and MapStreaming.aspx by following the flow of [25]. The Default.aspx form contains all the controls required to perform various operations on the map and geospatial data. This form is populated with following controls:

- cblLayers (CheckBoxList) for selection of layers to be displayed
- ddlLayers (DropDownList) for selection of layer to be identified
- litIdentifyResult (Literal) for display the result of identified map features
- ibtnMap (ImageButton) for display the map
- ibtnZoomIn (ImageButton) to perform zoom in
- ibtnZoomOut (ImageButton) to perform zoom out
- ibtnPan (ImageButton) to perform panning
- ibtnIdentify (ImageButton) to identify the features of selected layer
- ibtnFullExtent (ImageButton) to display the full extent map
- ibtnRefresh (ImageButton) to refresh the map
- rblAction (RadioButtonList) to set the map action with item values Zoom In, Zoom Out, Pan and Identify The MapStreaming.aspx form does not contain any control and is used to send the image stream to the ibtnMap control. The following Figure 7 represents the flow of code among the components of prototype application.

The next section describes the code to implement the essential functions of mapping (detailed code is available on request from authors).

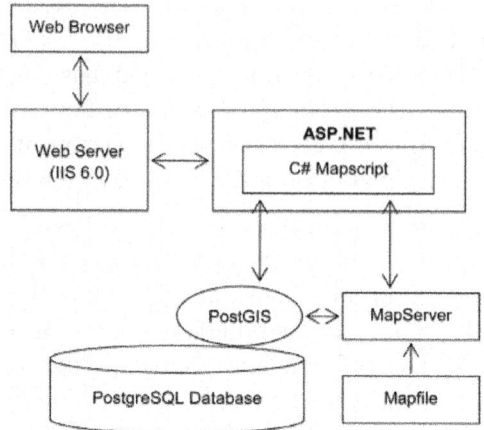

Figure 6: Architecture of prototype mapping application.

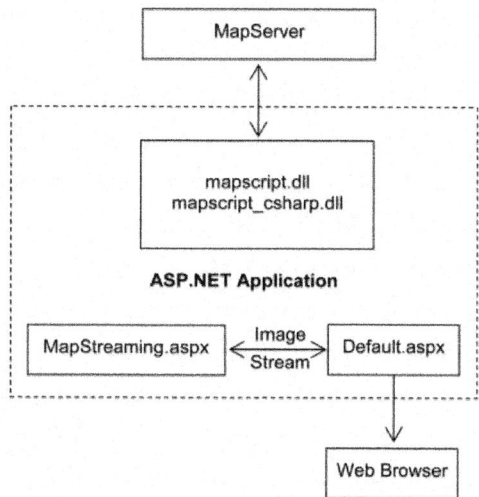

Figure 7: Flow of code [25].

C#.NET Mapscript for "Default.aspx"

using System;

using System.Web.UI;

using OSGeo.MapServer; //For MapScript public partial class _Default : System.Web.UI.Page

{

```
mapObj map;
protected void Page_Load(object sender, EventArgs e)
{
if (!Page.IsPostBack)//First Access to Map
{
//Image Stream from MapServer ibtnMap.ImageUrl = "MapStreaming.
aspx?MAPACTION =INITMAP";
mapObj map = new mapObj(Server.MapPath("App_Data/ MapFile.map"));
for (int i = 0; i < map.numlayers; i++)
{
layerObj layer = map.getLayer(i);
if (layer.status == (int)mapscript.MS_ON) {
cblLayers.Items[i].Selected = true;
}
}
}
else { //Next Access to Map (By Session)
map = (mapObj)Session["MAP"];
}
}
protected void ibtnMap_Click(object sender, ImageClick EventArgs e)
{
lblInfo.Text = "";
string mapAction = "";
string activeLayer = ddlLayers.SelectedValue;
//Check and Set Map Action switch (rblAction.SelectedItem.Text.ToUpper())
{
case "ZOOM IN":
mapAction = "ZOOMIN";
break;
case "ZOOM OUT":
mapAction = "ZOOMOUT";
```

```
break;
case "PAN":
mapAction = "PAN";
break;
case "IDENTIFY":
mapAction = "IDENTIFY";
break;
}
```

//Image Stream from MapServer to ibtnMap according to required Map Action
ibtnMap.ImageUrl = "MapStreaming.aspx?MAPACTION =" + mapAction +
"&X=" + e.X + "&Y=" + e.Y + "&ACTIVELAYER=" + activeLayer;
```
}
```

//Setting rblAction for different Map Operations protected void ibtnZoomIn_
Click(object sender, Image ClickEventArgs e) {

rblAction.SelectedValue = "Zoom In";
```
}
```

protected void ibtnZoomOut_Click(object sender, Image ClickEventArgs e) {
rblAction.SelectedValue = "Zoom Out";
```
}
```

protected void ibtnPan_Click(object sender, ImageClick EventArgs e) {
rblAction.SelectedValue = "Pan";
```
}
```

protected void ibtnIdentify_Click(object sender, Image ClickEventArgs e) {
rblAction.SelectedValue = "Identify";
```
}
```

//For Full Extent Map protected void ibtnFullExtent_Click(object sender,
Image ClickEventArgs e) {

ibtnMap.ImageUrl = "MapStreaming.aspx?MAPACTION =FULLEXTENT";
```
}
```

//For Refreshing the Map according to Selected (cblLayers Checked) Map
Layers protected void ibtnRefresh_Click(object sender, Image ClickEventArgs
e)
```
{
```

```
for (int i = 0; i < cblLayers.Items.Count; i++)
{
layerObj layer = map.getLayerByName(cblLayers.Items[i]. Value);
if (cblLayers.Items[i].Selected) {
layer.status = (int)mapscript.MS_ON;
}
else {
layer.status = (int)mapscript.MS_OFF;
}
}
ibtnMap.ImageUrl = "MapStreaming.aspx?MAPACTION =REFRESHMAP";
}
}
```

C#.NET Mapscript for "MapStreaming.aspx"

```
using System;
using System.IO;
using System.Web.UI;
using System.Drawing;
using OSGeo.MapServer; //For MapScript public partial class MapStreaming
: System.Web.UI.Page
{
mapObj map;
rectObj originalExtent;
private enum zoomMode {zoomIn = 0, zoomOut=1} //Enumerator for
Zooming protected void Page_Load(object sender, EventArgs e)
{
//Read Map from Session (if exist), Otherwhise Create new using Mapfile map
= (mapObj)Session["MAP"];
if (map == null)
{
map = new mapObj(Server.MapPath("App_Data/Map File. map"));
originalExtent = new rectObj(map.extent.minx, map.extent.miny, map.extent.
maxx, map.extent.maxy, 0);
```

```
Session["ORIGINALEXTENT"] = originalExtent;
}
//Variables to read the required Map Action, (x,y) position and active Layer
String mapAction;
Double x = 0; //Longitude Double y = 0; //Latitude String activeLayer;
//Read Map Action from QueryString mapAction = Request.
QueryString["MAPACTION"]. ToString().ToUpper();
//Read (x,y) position from QueryString if (Request.QueryString["X"] != null
&& Request.Query String["Y"] != null) {
x = Double.Parse(Request.QueryString["X"].ToString());
y = Double.Parse(Request.QueryString["Y"].ToString());
}
//Read active Layer from QueryString if (Request.
QueryString["ACTIVELAYER"] != null) {
activeLayer = Request.QueryString["ACTIVELAYER"]. ToString();
}
//Call for different methods based on Map Action switch (mapAction)
{
case "ZOOMIN":
doZoom(zoomMode.zoomIn, x, y); // Zoom In break;
case "ZOOMOUT":
doZoom(zoomMode.zoomOut, x, y); // Zoom Out break;
case "PAN":
doPan(x,y); // Panning break;
case "IDENTIFY":
doIdentify(x, y, activeLayer); // Feature Identification break;
case "FULLEXTENT":
doFullExtent(); //Full Extent break;
case "REFRESHMAP":
break;
}
doRefresh();
Session["MAP"] = map;
```

}
private void doZoom(zoomMode zMode, Double x, Double y) { }
private void doPan(Double x, Double y) {   }
private void doIdentify(Double x, Double y, String active Layer) { }
private void doFullExtent() {   }
private void doRefresh() {   }

CONCLUSION

In any project, for the developing countries like India, cost is the primary consideration for the adaptability of spatial technology. Under these circumstances, free and open-source geospatial tools provide all the data storage, analysis and information visualization for free. This paper provides a foundation in web-based mapping using open-source geospatial tools and web technology. It highlights building a prototype mapping application by providing step-by-step instructions which can encourage the eager developers and interested readers to publish their maps on the web with no prior technical experience in map servers. The whole prototype lives in a common web-based application implemented in ASP.NET with C#.NET MapScript exploiting MapServer for mapping functionalities and PostGIS for its connection with the PostgreSQL databse. The techniques and source code of this research work can be applied as the laboratory exercise or a course curriculum and can help a large number of organizations to publish their geographic information on the web, especially those small organizations that cannot afford proprietary software's cost, complexity, training cost and special requirements.

ACKNOWLEDGEMENTS

The corresponding author is thankful to University Grants Commission, India for providing financial support in the form of research fellowship under RFSMS scheme.

REFERENCES

1.   R. Mari, L. Bottai, C. Busillo, F. Calastrini, B. Gozzini and G. Gualtieri, "A GIS-Based Interactive Web Decision Support System for Planning Wind Farms in Tuscany (Italy)," Renewable Energy, Vol. 36, No. 2, 2011, pp. 754- 763.http://dx.doi.org/10.1016/j.renene.2010.07.005

2.   M. Ozimec, M. Natter and T. Reutterer, "Geographical Information Systems-Based Marketing Decisions: Effects of Alternative Visualizations on Decision Quality," Journal of Marketing, Vol. 74, No. 6, 2010, pp. 94-

110. http://dx.doi.org/10.1509/jmkg.74.6.94

3.   Okabe and K. Okunuki, "A Computational Method for Estimating the Demand of Retail Stores on a Street Network and Its Implementation in GIS," Transactions in GIS, Vol. 5, No. 3, 2001, pp. 209-220. http://dx.doi.org/10.1111/1467-9671.00078

4.   K. Ziliaskopoulos and S. T. Waller, "An InternetBased Geographic Information System That Integrates Data, Models and Users for Transportation Applications," Transportation Research Part C, Vol. 8, No. 1-6, 2000, pp. 427-444.

5.   F. Xie, "Design and Implementation of Highway Management System Based WebGIS," Journal of Networks, Vol. 5, No. 12, 2010, pp. 1389-1392.http://dx.doi.org/10.4304/jnw.5.12.1389-1392

6.   S. P. Singh, J. Sharma and P. Singh, "A Geo-Referenced Information System for Tourism (GeoRIST)," International Journal of Geomatics and Geosciences, Vol. 2, No. 2, 2011, pp. 456-464.

7.   P. S. Singh, D. Chutiya and S. Sudhakar, "Development of a Web Based GIS Application for Spatial Natural Resources Information System Using Effective Open Source Software and Standards," Journal of Geographic Information System, Vol. 4, No. 3, 2012, pp. 261-266. http://dx.doi.org/10.4236/jgis.2012.43031

8.   T. Q. Zeng and Q. Zhou, "Optimal Spatial Decision Making Using GIS: A Prototype of a Real Estate Geographical Information System (REGIS)," International Journal of Geographical Information Science, Vol. 15, No. 4, 2001, pp. 307-321.http://dx.doi.org/10.1080/136588101300304034

9.   T. Oswari, E. S. Suhendra, E. Haryatmi and F. Agustina, "Prototype Geographic Information Systems Mapping of Crop Products Featured Local," Journal of Geographic Information System, Vol. 5, No. 3, 2013, pp. 193-197.http://dx.doi.org/10.4236/jgis.2013.53018

10. Evans and C. E. Sabel, "Open-Source Web-Based Geographical Information System for Health Exposure Assessment," International Journal of Health Geographics, Vol. 11, 2012, p. 2.

11. R. M. Mullner, K. Chung, K. G. Croke and E. K. Menash, "Geographic Information Systems in Public Health and Medicine," Journal of Medical Systems, Vol. 28, No. 3, 2004, pp. 215-221. http://dx.doi.org/10.1023/B:JOMS.0000032972.29060.dd

12. Boyles, "GIS Means Business," ESRI, Redlands, 2002.

13. M. N. K. Boulos and K. Honda, "Web GIS in Practice IV: Publishing Your Health Maps and Connecting to Remote WMS Sources Using the Open

Source UMN MapServer and DM Solutions MapLab," International Journal of Health Geographics, Vol. 5, 2006, p. 6.

14. Q. Yi, R. E. Hoskins, E. A. Hillringhouse, S. S. Sorensen, M. W. Oberle, S. S. Fuller and J. C. Wallace, "Integrating Open-Source Technologies to Build Low-Cost Information Systems for Improved Access to Public Health Data," International Journal of Health Geographics, Vol. 7, 2008, p. 29.

15. S. Steiniger and E. Bocher, "An Overview on Current Free and Open Source Desktop GIS Developments," International Journal of Geographical Information Science, Vol. 23, No. 10, 2009, pp. 1345-1370. http://dx.doi.org/10.1080/13658810802634956

16. J. Lowe, "Spatial on a Shoestring: Leveraging free Open Source Software," Geospatial Solutions, Vol. 12, No. 6, 2002, pp. 42-45.

17. T. Mitchell, "Open Source Geo Tools," O'Reilly Where Where 2.0 Conference, San Francisco, 29-30 June 2005.

18. T. Mitchell, "An Introduction to Open Source Geospatial Tools," 2005. http://www.oreillynet.com/pub/a/network/2005/06/10/osgeospatial.html

19. W. M. Lee, "C#. NET Web Developers Guide," Syngress, Rockland, 2002.

20. Mansourian, A. Fasihi and M. Taleai, "A Web-Based Spatial Decision Support System to Enhance Public Participation in Urban Planning Processes," Journal of Spatial Science, Vol. 56, No. 2, 2001, pp. 269-282. http://dx.doi.org/10.1080/14498596.2011.623347

21. G. Othman, B. Mohamed, A. Bahauddin, A. P. M. Som and S. I. Omar, "A Geographic Information System Based Approach for Mapping Tourist Accommodations in the East Coast States of Malaysia," World Applied Sciences Journal, Vol. 10, 2010, pp. 14-23.

22. PostGIS. http://postgis.net/

23. S. Steiniger and A. J. Hunter, "Free and Open Source GIS Software for Building a Spatial Data Infrastructure," In: E. Bocher and M. Neteler, Eds., Geospatial Free and Open Source Software in the 21st Century, Springer-Verlag Berlin Heidelberg, 2012, pp. 247-261.

24. P. Ramsey, "The State of Open Source GIS," 2007.http://www.refractions.net/expertise/whitepapers/opensourcesurvey/survey-open-source-2007-12.pdf

25. P. Corti, "Thinking in GIS," 2006. http://www.paolocorti.net/2006/09/20/mapserver-tutorial-for-c-mapscript-asp-net/

Chapter 4

GEOSTATISTICAL APPROACH FOR SITE SUITABILITY MAPPING OF DEGRADED MANGROVE FOREST IN THE MAHAKAM DELTA, INDONESIA

Ali Suhardiman[1,2], Satoshi Tsuyuki[1], Muhammad Sumaryono[2], Yohanes Budi Sulistioadi[2,3]

[1]Department of Global Agricultural Sciences, The University of Tokyo, Tokyo, Japan

[2]Department of Forest Science, University of Mulawarman, Samarinda, Indonesia

[3]Division of Geodetic Science, School of Earth Science, The Ohio State University, Columbus, USA

ABSTRACT

As part of operational guidance of mangrove forest rehabilitation in the Mahakam delta, Indonesia, site suitability mapping for 14 species of mangrove was modelled by combining 4 underlying factors—clay, sand, salinity and tidal inundation. Semivariogram analysis and a geographic information system (GIS) were used to apply a site-suitability model, while kriging interpolation generated surface layers, based on sample point data collection. The tidal inundation map was derived from a tide table and a digital elevation model from topographic maps. The final site-suitability maps were produced using spatial analysis technique, by overlaying all surface layers. We used a Gaussian model to adjust a semivariogram graph in order to help to understand the variation of sample data values, and create a natural surface layer of data distribution over the area of study. By examining the statistical value and the visual inspection of surface layers, we saw that the models were consistent with the expected data behavior; therefore, we assumed that interpolation has been carried out appropriately. Our site-suitability map showed that Avicennia species was the most suitable species and matched with 50% of the study area, followed by Nypa fruticans, which occupied about 42%. These results were actually consistent with the mangrove zoning pattern in the region prior to deforestation and conversion.

INTRODUCTION

The Mahakam delta is located at the mouth of the Mahakam river on the east coast of Kalimantan, Indonesia. Including the distributary and interdistributary channels, the delta plain covers an area of 1500 km². It was originally covered by dense mangrove, composed of mainly Avicennia, Sonneratia, Rhizophora, and Nypa species, and fresh water mangroves [1]. This ecosystem was disturbed since the introduction of shrimp farming. Human settlements also expanded because of the oil and gas exploration and exploitation, which attracted both workers and land speculators [2]. This ecological disturbance reached its peak during Asian economic crisis (1997- 2000) when shrimp price on the global market inflated to 4 - 5 times the normal level due to devaluation of Indonesian rupiah against international currency. This triggered mangrove forest conversion to ponds, involving heavy equipment, on a massive scale [3].

In 2001, 75% of the delta had been changed to ponds [4], which subsequently reduced to 52% in 2007, as reported by the Fishery and Oceanic Affairs Office of the Kutai Kartanegara District government [5]. Almost 90% of the Mahakam delta region belongs to the Indonesian Ministry of Forestry and is classified to be a production forest, however three other different government sectors lay claim to the management of the Mahakam delta: the Fishery Department, Interior Affairs Department, and the Environment Department [3]. Due to unclear management, laws and regulations, land use control of this area failed to be enforced.

Massive degradation of mangrove ecosystems has prompted a world-wide movement to plant new areas of mangroves [6]. Unfortunately, mangrove rehabilitation has often been carried out simply by planting mangrove seedlings, without adequate site assessment, or subsequent evaluation of the success of planting at the ecosystem level [7,8]. In this paper, we applied geostatistical and geographical information system (GIS) to generate site-suitability mapping for mangrove rehabilitation guidance in the Mahakam delta.

MATERIALS AND METHODS

Study Area

The Mahakam delta is located between 117.30°E - 117.62°E and 0.49°S - 0.82°S (Figure 1). It comprises 46 sedimentation-formed islands [9]. For this study, we focused on areas (approximately 40,636 ha) where sample points were distributed relatively balance in term of spatial representation and therefore, data interpolation would give the best prediction.

Model Criteria and Flow Chart of the Study

In this study, we employed the technical guidance on mangrove plantation issued by Indonesian Ministry of Forestry in 2004. Geographical information system model was developed using the criteria presented in Table 1. ArcGIS 10 embedded geostatistical analysis tool was used to apply the model. Data normality were examined prior to geostatistics analysis. Geostatistical analysis tool was used to perform semivariogram analysis, ordinary kriging interpolation and subsequent cross-validation. Ordinary kriging interpolation produced a surface layer (continuous data) of all the data from our sample points. By superimposing all related surface layers, we can produce map of site suitability for mangrove rehabilitation over the entire area of the study. The major analysis steps used in this study were shown in Figure 2.

The physical data collected were soil texture (percentage of clay and sand fraction) and salinity. Tidal inundation data was derived from combination of a digital elevation model (DEM) and the 2009 tide table, obtained from the hydro-oceanographic division of the Indonesian navy. The digital elevation model was derived from the height points of the Mahakam delta topographic maps.

The tide table contained 12 months of hourly sea level rise predictions for a specific area. January, April, July and October were selected to represent the 2009 tidal prediction. We assumed each selected month would capture the seasonal effect of tidal dynamics in the area of interest. The East Kalimantan province is characterized by a tropical rain forest climate with a dry (May to September) and a wet (October to April) season [10]. The dry and wet seasons were represented by the months of July and January, respectively, while April and October represented the transitional months.

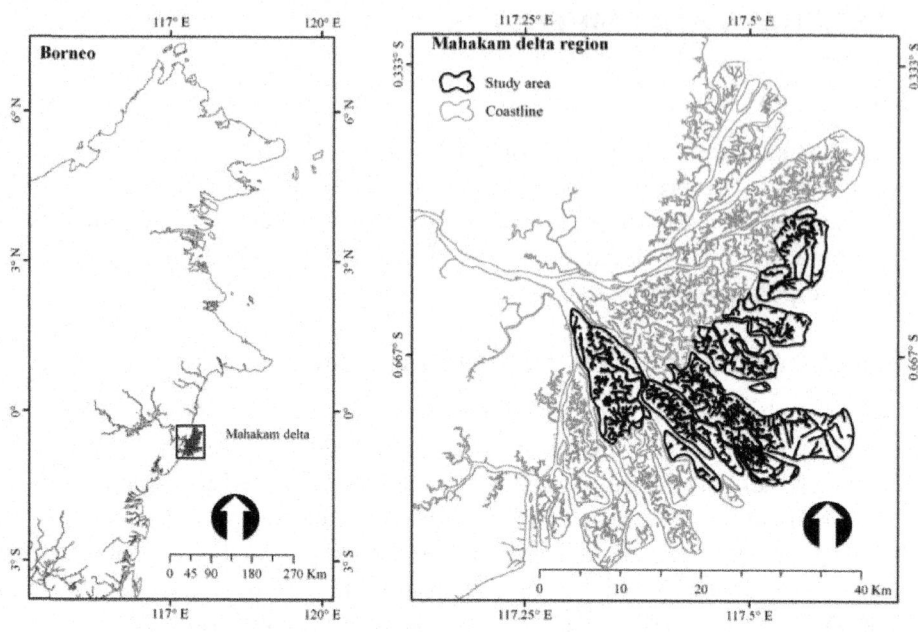

Figure 1: Location of Mahakam delta in the east coast of Borneo Island. The area sat the focus of this study are outlined by bold lines.

Table 1: Technical guidance of mangrove plantation in the forested areas of Indonesia

Zones	Species	Salinity (ppt)	Sandy fraction tolerance	Mud tolerance	Frequency of tidal inundation
A	*Rhizophora mucronata* Lamk. *Rhizophora apiculata* Bl.	10 - 30	Moderate	Suitable	<20 days/month
B	*Rhizophora stylosa* Griff *Sonneratia alba* J.Smith.	10 - 30	Suitable	Suitable	<20 days/month
C	*Bruguiera parviflora* (Roxb.) Wight & Arn. ex Griffith. *Bruguiera sexangula* (Lour.) Poir.	10 - 30	Moderate	Suitable	10 - 20 days/month
D	*Bruguiera gymnorrhiza* (L.) Savigny.	10 - 30	Not suitable	Moderate	10 - 20 days/month
E	*Sonneratia caseolaris* (L.) Engl.	10 - 30	Moderate	Moderate	<20 days/month
F	*Xylocarpus granatum* J.Koenig. *Heritiera littoralis* Dryand.	10 - 30	Moderate	Moderate	>10 days/month
G	*Lumnitzera racemosa* Willd.	10 - 30	Suitable	Moderate	>10 days/month
H	*Cerbera manghas* L.	0 - 10	Moderate	Moderate	>10 days/month
I	*Nypa fruticans* Wurmb.	0 - 10	Not suitable	Suitable	<20 days/month
J	*Avicennia* sp.	10 - 30	Not suitable	Suitable	<20 days/month

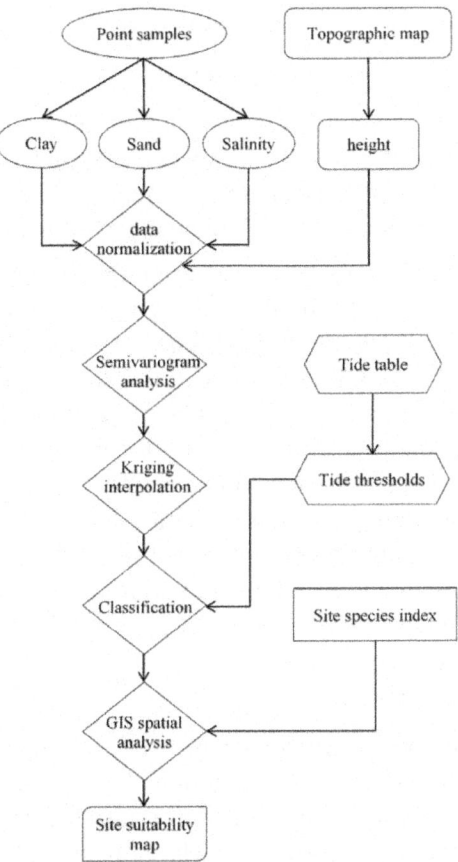

Figure 2: The flow chart of the study, showing the major steps and the site-suitability map as the final result.

Semivariogram

A semivariogram is based on an assumption that properties of regions in proximity to one another are more alike than those farther away. In order to get a better interpolation result based on semivariogram analysis, we needed normally distributed data. The Shapiro-Wilk test was used to quantify the normality of the data. We used the BoxCox and logarithmic transformation to normalize the data. After that, we used the semivariogram analysis to quantify the spatial relationship between the sample points. The semivariogram symbolized by g(h) is the average variance between observations separated by a distance h. It was computed using formula:

$$\gamma(h) = \frac{1}{2N(h)} \sum_{i=1}^{N(h)} \{z(x_i) - z(x_i + h)\}^2$$

(1)

where $\gamma(h)$ is the experimental semivariogram value at a distance h; N(h) is number of sample value pairs within distance h; and $z(x_i)$, $z(x_i + h)$ are the sample values at two points separated by distance h. All pairs of points separated by distance h (lagh) were used to calculate the experimental semivariogram.

The semivariogram parameters are the range, total sill and nugget ratio. The total sill is the sum of the sill and nugget, while the nugget ratio is the ratio of the nugget to the total sill [11]. Lag is the distance between two pairs data; the sill is the maximum level of $\gamma(h)$; the range is the point on the h axis where $\gamma(h)$ reaches a maximum [12]. The experimental semivariogram will be fitted using a mathematical model used for interpolation. There are many models available, but the most acceptable ones for continuous variables would be spherical, exponential, Gaussian or combinations of these models [12]. In this study, we used a Gaussian model with a small range and sill value, which means this model predicted smaller data variability compared to the spherical and exponential models.

Kriging Interpolation

We applied Kriging interpolation to clay, sand, salinity and height point data following the semivariogram investigation. Reference [13] discovered that kriging interpolation was better suited for prediction and mapping of the distribution of chemical soil properties while reference [14] found that kriging interpolation had successfully explained soil properties. We chose widely used ordinary kriging interpolation to generate the model, assuming the absence of non-linear trends and focusing only on the spatially correlated component [15].

The characteristics of an interpolated surface layer can be controlled by limiting the input points used in the calculation of the output cell values. Specifying the maximum and minimum number of points to be sampled will return the points closest to the output cell location, until the maximum number is reached. This number is then used in calculating output values. In this study, we used a combination of maximum and minimum number of neighboring points to be 5 and 2, respectively. Another method to control the interpolated surface is by defining either a circular or elliptical model to enclose the points that are used to predict output cell values. Furthermore, these circular or elliptical models could be divided into sectors. We used a circular model split by a diagonal cross section (4 sectors with a 45° offset) for calculating output cell values.

Surface layer-based interpolation data were classified according to criteria shown in Table 1. To divide the surface layer into classes (suitable, moderate or not suitable) especially for clay and sand data, we used Jenks natural breaks provided in most of GIS softwares. Salinity data was divided into two classes: 0 - 10 ppt (part per thousand) and 10 - 30 ppt. From DEM data, we defined three classes of tidal inundation based on the number of inundation days in a month: 1) more than 20 days in a month; 2) between 10 - 20 days; and 3) less than 10 days in a month.

Validation

Cross validation of sample points was used to measure the model accuracy. The following statistics were used:

root mean square (RMS), which measured error size sensitive to outliers [16], and standardized RMS [15]. RMS was calculated using the following formula:

$$\text{RMS} = \sqrt{\frac{1}{n}\sum_{i=1}^{n}\left(Z_{i,act} - Z_{i,est}\right)^2}$$

(2)

By dividing RMS with standard error then standardized RMS would be obtained. Here the variables are defined as follows: $Z_{i,act}$ = the known value at point i $Z_{i,est}$ = the estimated value at point i n = the number of points.

RESULTS

Normalizing Data

The normality test using Shapiro-Wilk method indicated that our data was not normally distributed (Table 2). Therefore, data normalization was imperative prior to kriging interpolation. Only clay data appeared to be normally distributed, since its p-value was close to 0.05. We performed a standard statistical calculation and found that sand, salinity, and height data exhibited higher variation than the clay data. Coincidentally, the high variation was related to the p-value of Shapiro-Wilk. We used the Box-Cox transformation for all data, except clay, which was transformed using a logarithmic method. The Box-Cox improved the efficacy of normalizing and variance, equalizing for both positivelyand negatively-skewed variables [17].

Producing the Tidal Inundation Map

We analyzed 4 monthly tide tables and found that the tide rose somewhere between 0 - 3 m. The lowest altitude area was assumed to be inundated in longer period than higher altitude area. We summed up days exceeding a certain tidal height interval (0.1 m) and determined as inundation period starts from 0 m above sea level. Following tidal inundation classes (Table 1), we obtained specific height thresholds to separate the classes that is 1.1 and 1.6 m of altitude (Figure 3).

Semivariogram and Kriging Interpolation Analysis

Lag size and the number of lags are two parameters that contribute to the number of pair points generated in a semivariogram graph. The number of lags was set to 12, which is the default value set by the geostatistical analysis tool. From Figure 4, we found that variation in the sand data was confirmed to be the highest one, as indicated by the nugget value equal to 16.024, as shown in Table 3. However, we also found that salinity data need a longer distance to reach the sill value (nearly 45 km).

Table 2: Descriptive statistics of the four types of data

No.	Properties	Number	Mean	Coefficient of Variation	Skewness	Kurtosis	Shapiro-Wilks (p-value)
1.	Clay	66	34.97	0.57	0.44	−0.61	0.045
2.	Sand	66	35.01	0.83	0.53	−1.05	7.561×10^{-5}
3.	Salinity	67	12.64	0.92	0.46	−1.14	1.939×10^{-5}
4.	Height	641	1.434	0.83	0.58	0.95	3.763×10^{-24}

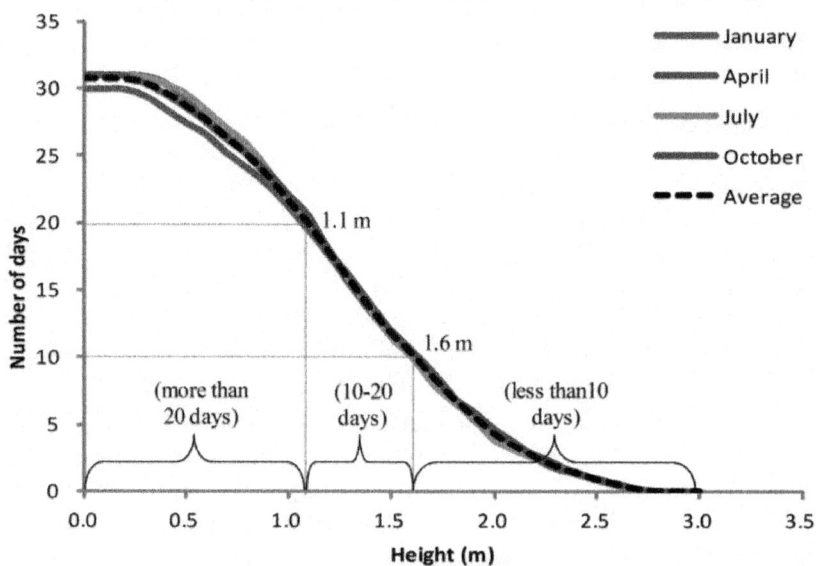

Figure 3: The number of inundation days for each of the height levels summarized from tide table.

Kriging interpolation accuracy was inspected through RMS and standardized RMS. However we could not compared RMS values, since the units were different. From the standardized RMS value, we found that clay and height data underestimated the variability of prediction (standardized RMS < 1), while sand and salinity would be overestimated (standardized RMS > 1).

Surface Classification Using GIS

Kriging interpolation produced surface layers, as shown in Figure 5. Clay fraction data after interpolation ranged between 7.4% and 52.8%. Two clay classes were defined using the Jenks natural break. Moderate class for clay was within the range 7.4% - 28.2%, while the suitable class was 28.2% - 52.8%. Sand data after interpolation ranged between 5.5% and 76.0%. Using similar method, three classes were generated and classified as: not suitable (5.5% - 25.6%), moderate (25.6% - 46.8%) and suitable (46.8% - 76.0%). Digital elevation model was classified as 1) elevation less than 1.1 m, which would be inundated more than 20 days a month; 2) elevation higher than 1.6 m, which would be inundated less than 10 days in a month; and 3) elevation ranging between 1.1 and 1.6 m, which would be inundated within 10 - 20 days in a month.

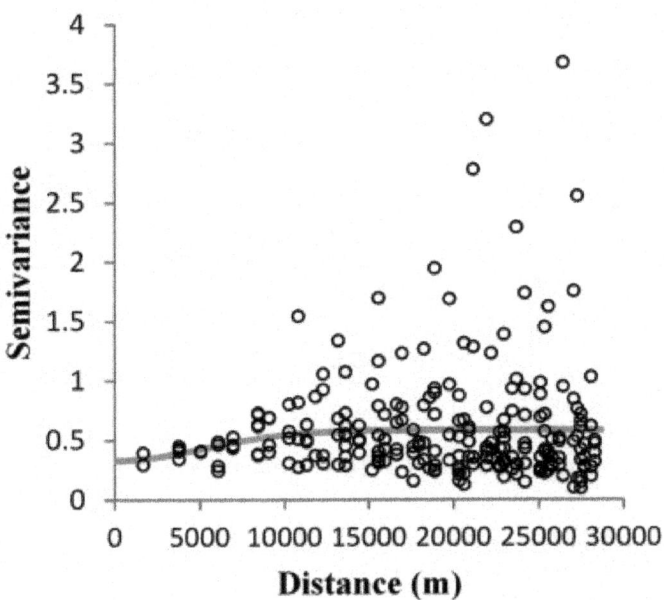

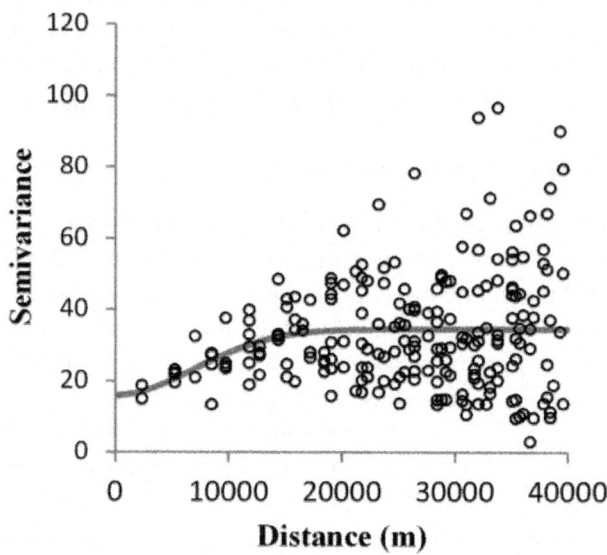

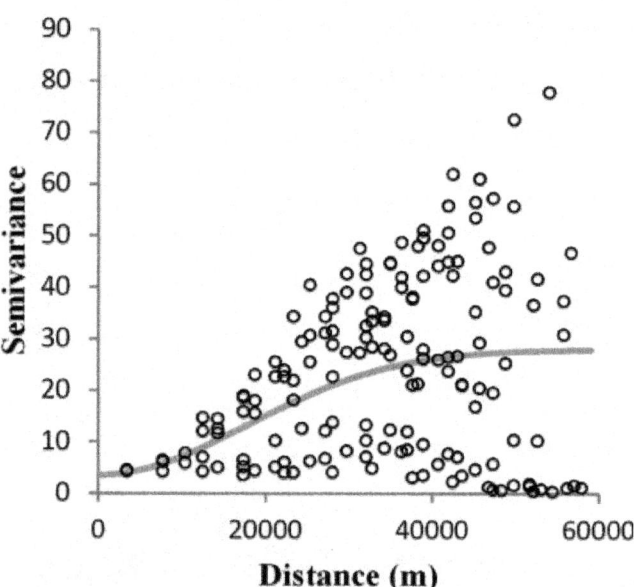

Semivariogram of height data

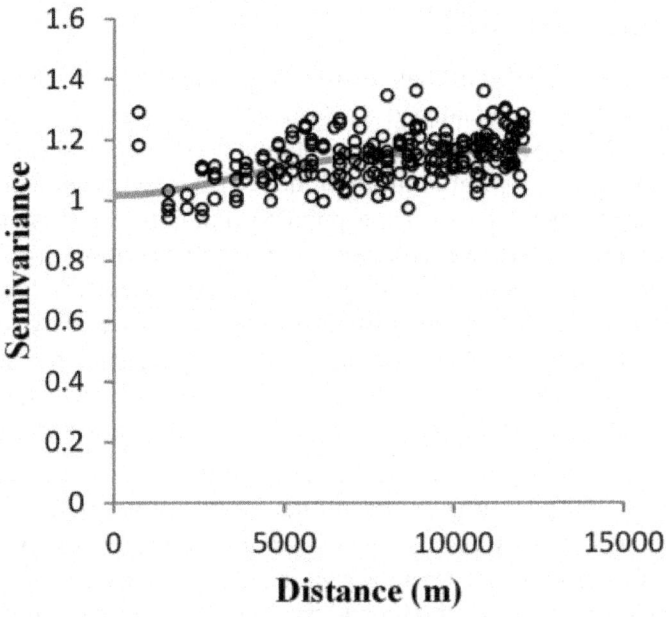

Figure 4: Omnidirectional semivariogram fitted by Gaussian model for all data.

Table 3: Semivariogram parameters for each data with RMS and standardized RMS values after kriging interpolation

Parameters	Semivariogram				Kriging interpolation	
	Range	Nugget	Total sill	Nugget ratio	RMS error	Standardized RMS
Clay	12693.267	0.331	0.584	0.567	21.821	0.779
Sand	17035.230	16.024	34.510	0.464	27.010	1.186
Salinity	43432.550	3.579	28.034	0.021	6.284	1.047
Height	8703.099	1.017	1.163	0.874	1.086	0.878

According to Table 1, there are ten unique zones of mangrove species. Some of the species shared similar site characteristics, such as Rhizophora mucronata and R. apiculata or Bruguiera parviflora and B. sexangula. The GIS-based overlay of whole surface layers produced 12 unique sites with different combination of all data. A site-suitability map of mangrove species was generated corresponding to criteria in Table 1. The results showed that three site zones were not represented in any GISunique sites, i.e. zone B (Rhizophora stylosa and Sonneratia alba), G (Lumnitzera racemosa), and H (Cerbera manghas). The final map of site suitability of mangrove species is presented in Figure 6.

DISCUSSION

Geostatistical Analysis

Reference [18] explained that non-normal data is actually common in geostatistical data, especially if the observed data is discrete, i.e. ppm, percentages of grade, etc. Moreover, reference [19] showed that in the case of soils, they naturally form in different depositional environments; therefore, their physical properties vary from point to point (in the horizontal as well as vertical plane). Thus, we assumed the existing variation of our data were natural. Semivariogram graphs shown in Figure 3 describe the relation of semivariance value (y-axis) and distances (xaxis). All data (clay, sand, salinity and height) seemed to follow the theoretical assumption of semivariogram, where points that are close to each other are expected to have small variance, and points that area farther apart are expected to have higher variance [15].

As shown in Table 3, the highest value for the sill element was that of sand, which means that the variance among the sand sample point data is greater than for other data. However, semivariogram measuring salinity showed that more than 43 km were needed to get variance to decrease, while clay had the smallest such distance. Sand data had the highest nugget value, which might indicate the highest measurement errors. Nevertheless, all semivariogram graphs showed a natural experimental omnivariogram.

Clay data had RMS error equal to 21.82%, which was smaller than sand with 27.01%. While salinity had RMS error of 6.28 ppt and height error 1.09 m. Standardized RMS value had juxtaposed our data into 2 groups (clay with height and sand with salinity) respecting to the estimation of the variability of the predictions. However, according to reference [20], statistical value is not the only source of assessing the interpolation quality; we also need to carefully study the visual quality of the surface model generated by kriging interpolation to detect distinct spatial patterns and visual faithfulness.

Analysis of Spatial Data Interpolation

We compiled 66 sample points for soil texture and 67 sample points for salinity, both onand off-shore. Samples were not spread evenly across the area but rather clumped in the southern part and missing in the central part. As a result, the study only focused on some parts of the Mahakam delta, which would have the best interpolation results. The interpolation process ignored the distributary channel and assumed that the topographic features in the delta plain were negligible [21]. Soil type in the Mahakam delta is loamy, varying from silt loam to clay loam, except for the coastline which is loam with sand [2]. According to Figure 5, we found consistency in this fact, whereas higher clay concentration dominate over the area of study. The sand fraction concentration is located at the outer delta, while clay mostly deposits in the center of delta, straight to the Mahakam river flow discharge direction. Although pond construction altered land, however, soil texture was assumed to remain fairly constant and was unchanged by management practices [22,23]. A suitable class for clay and sand means that both fractions have a high percentage. A high concentration of clay dominated 95% of the study area. A species preference for high percentage of clay implies that species was intolerant to the high percentage of sand and vice versa. Salinity also showed a natural distribution, where higher concentration situated in the outer delta rather than the inner delta. Salinity data is one of the sea water environmental factors that would spread gently and smoothly, especially in the estuary region. The total salinity range of the open ocean is about 30 - 40 parts per thousand (or ‰) while salinity in coastal estuaries can be much lower [24]. Altitude ranged from 0 - 4 m above sea level, with an average 2.26 m. A digital elevation model was generated

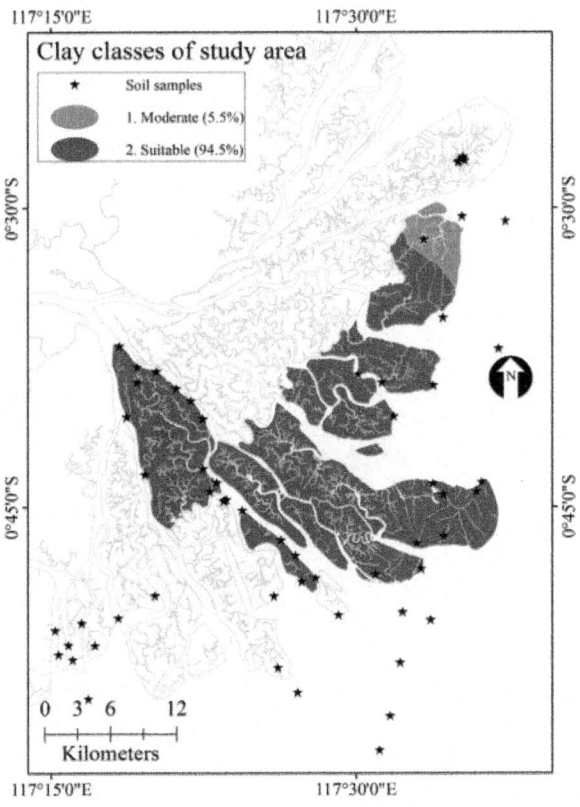

(a)

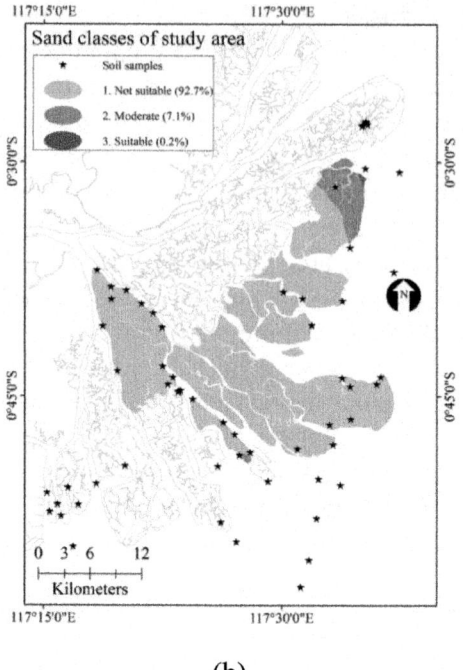

(b)

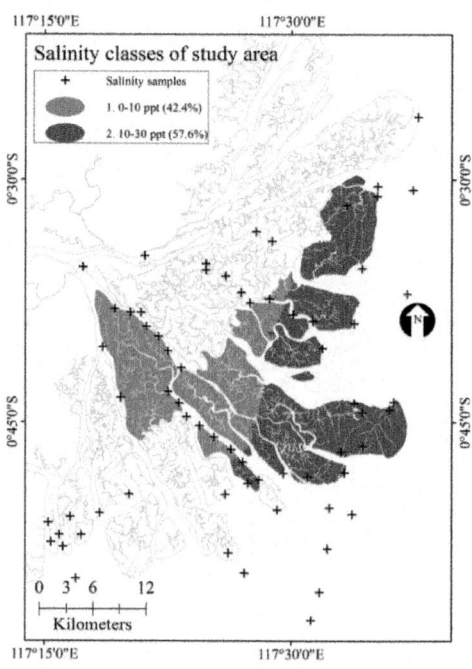

(c)

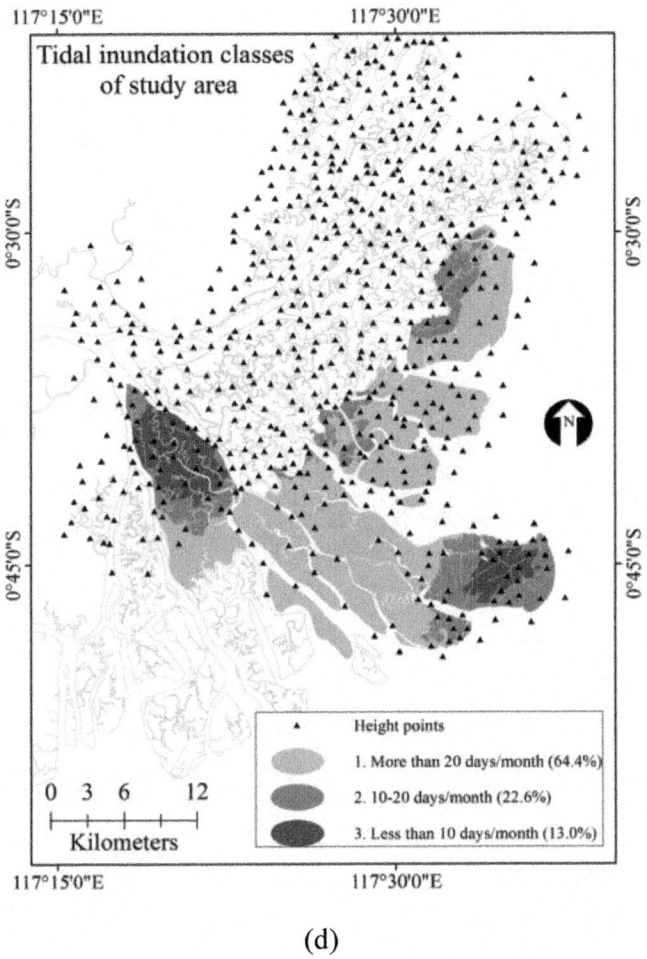

(d)

Figure 5: Classification of surface layers: (a) clay; (b) sand; (c) salinity; and (d) tidal inundation.

by kriging interpolation. The interpolation produced continuous elevation, ranging from 0 to 4.64 m. Normally higher elevation area are located in the inner delta, but interpolation found another high elevation area in the outer delta, near the coastline (Figure 5(d)). This area used to be an agricultural spot planted with coconut trees. However, during the introduction of shrimp pond, this area was also converted. The construction of ponds changed the terrain to flat.

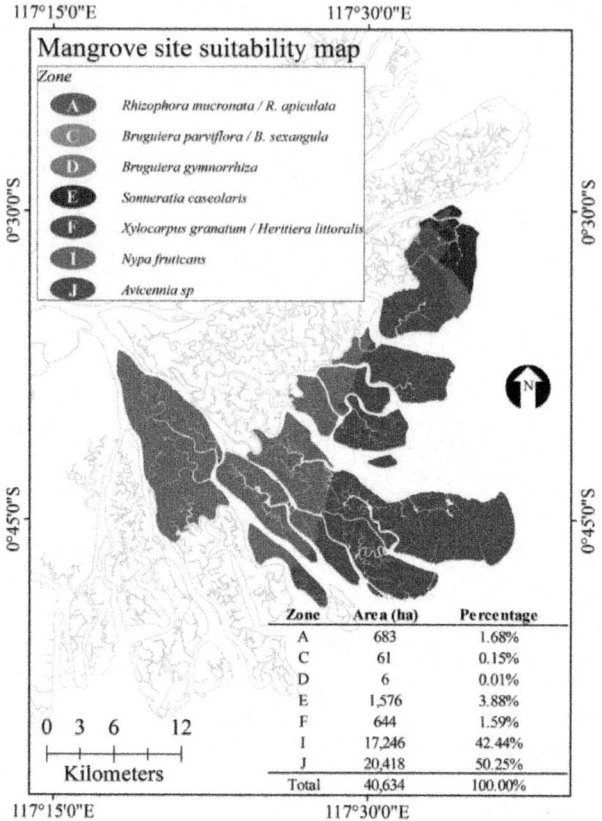

Figure 6: Site suitability map of our study sites generated using geostatistical analysis and GIS operations.

Site Suitability Mapping

Site suitability map as an output of spatial combination of four physical characteristics showed that theoretically 10 mangrove species were suitable for planting in the barren area or abandoned ponds. We calculated that 50.25% of the study area was suitable for Avicennia sp. and the other 42.22% suitable for Nypa fruticans. These two species occupied 92.69% of the study area. Another 7.14% was shared among Sonneratia caseolaris, Rhizophora mucronata, R. apiculata, Xylocarpus granatum and Heritiera littoralis. Only 0.16% would be suitable for Bruguiera species, which is known to prefer to grow at the boundary between mangroves and the coastal ecosystem, in the middle and upper inundation areas [25]. These results were consistent with the mangrove zoning, where Avicennia sp. is naturally found adjacent to the sea, together with Sonneratia species [26]. Meanwhile, Nypa fruticans seemed suitable to areas behind Avicennia sp., where brackish to almost freshwater streams exist and tidal effects decline.

Regarding to the rehabilitation efforts, site suitability map will assist the authority to prepare and implement rehabilitation strategy in this area. However, further studies are needed to validate this conceptual model by establishing small-scale demonstration plots and planting various mangrove species, to periodically compare their growth and mortality. Even though Avicennia sp. and Nypa fruticans seem suitable for about 92% of the total study area; it does not mean that other species will not survive if planted. Reference [26] emphasized that overlapping between species or zones occurred at many sites.

Site Suitability Mapping

Reference [27] reported that Total E&P Indonesia (company developing oil and gas) have planted more than 3 million mangrove trees on cleared areas, mainly along the pipeline distributional network. However, on a broader scale, mangrove rehabilitation in the Mahakam delta is still a minor project, due to unclear land ownership. Most of the ponds are effectively "owned" by the people operating them, so that plantation programs cannot be implemented in all necessary locations. Rehabilitation projects needs to be prepared carefully otherwise they will trigger social conflict. As a top priority, reference [2] suggested restoring mangroves in the areas that are incompatible with pond production due to poor soil quality or required for coastal protection.

In the case of the Mahakam delta region, a social approach for rehabilitation of the "occupied" land is a prerequisite to any such rehabilitation program. It is complicated by a number of factors, including the lack of clarity in land status, which, in theory, mostly belongs to the state, as well as the claims of the local people. Although these claims are illegal, it is impossible to restore the land without compensation to its current users; this is the main obstacle for rehabilitation programs in the region. Spending government fund for compensation is currently only permitted for private, not public, land. Recently, some small-scale silvofishery projects funded by a private company and an NGO had been introduced to this site in the last 3 - 4 years. However, most of the degraded areas remain barren and need immediate action to bring back the ecological function of mangrove forests for future generations.

CONCLUSION

Our study demonstrated an original approach to producing a site-suitability map of mangrove for rehabilitation purposes by combining 4 factors (clay, sand, salinity and height) using geostatistics. We proposed a new approach to making tidal inundation maps. Based on this study, we suggest height points linked with tide tables to produce tidal inundation maps. Statistical values and visual inspection of surface layers as a result of kriging interpolation yielded consistent results therefore, we assumed that interpolation processes have been carried out appropriately.

ACKNOWLEDGEMENTS

We would like to appreciate and thank Total E&P Indonesia, Dr. Iwan Suyatna from the Faculty of Fishery and Marine Sciences, University of Mulawarman, and the Faculty of Forestry, University of Mulawarman, for kind support in providing various data related to this study

REFERENCES

1. P. Sandjatmiko, A. M. Rony, H. Tarumadevyanto, I. Suyatna, Y. B. Sulistioadi, I. Tjitradjaja, L. Adrianto and D. G. Bengen, "Delta Mahakam Dalam Ruang dan Waktu. Ekosistem, Sumberdaya dan Pengelolaannya," BPMIGAS—Total E&P Indonesie dan Institute of Natural & Regional Resources (INRR), Jakarta, 2006. (In Indonesian).

2. R. Bosma, A. S. Sidik, P. Van Zwieten, A. Aditya and L. Visser, "Challenges of a Transition to a Sustainably Managed Shrimp Culture Agro-ecosystem in the Mahakam Delta, East Kalimantan, Indonesia," Wetlands Ecology and Management, Vol. 20, No. 2, 2012, pp. 89-99. http://dx.doi.org/10.1007/s11273-011-9244-0

3. R. A. Bourgeois, A. Gouyon, F. Jesus, P. Levang, W. Langeraar, F. Rahmadani, E. Sudiono and Y. B. Sulistioadi, "A Socio Economic and Institutional Analysis of Mahakam Delta Stakeholders," Total-Fina Elf, Belgium, Final Report to Total-Fina Elf, Contract No. 501125/DKI/204, 2002, pp. 108.

4. P. A. M. Van Zwieten, A. S. Sidik, Noryadi and I. S. Abdunnur, "Aquatic Food Production in the Coastal Zone: Data-based Perceptions on the Trade-off between Mariculture and Fisheries Production of the Mahakam delta and Estuary, East Kalimantan, Indonesia," In: C.T. Hoanh, T. P. Tuong, J. W. Gowing and B. Hardy, Eds., Environmental and Livelihoods in Tropical Coastal Zones: Managing Agriculture-Fishery-Aquaculture Conflicts, CABI, Oxfordshire, 2006, pp. 219-236. http://dx.doi.org/10.1079/9781845931070.0219

5. Anonym, "A Report on Detailed Ponds Mapping of Mahakam delta Using High Resolution Satellite Imagery," Fishery and Oceanic office of Kutai Kartanegara district, East Kalimantan province, Tenggarong, 2007.

(In Indonesian).

6. C. D. Field, "Rehabilitation of Mangrove Ecosystems: An Overview," Marine Pollution Bulletin, Vol.37, No. 8-12, 1999, pp. 383-392. http://dx.doi.org/10.1016/S0025-326X(99)00106-X

7. J. O Bosire, F. Dahdouh-Guebas, M. Walton, B. I. Crona, R. R. Lewis III, C. Field, J. G. Kairo and N. Koedam, "Functionality of Restored Mangroves: A Review," Aquatic Botany, Vol. 89, No. 2, 2008, pp. 251-259. http://dx.doi.org/10.1016/j.aquabot.2008.03.010

8. H. Ren, S. Jian, H. Lu, Q. Zhang, W. Shen, W. Han, Z. Yin and Q. Guo, "Restoration of Mangrove Plantations and Colonization by Native Species in Leizhou Bay, South China," Ecol Res, Vol. 23, No. 2, 2008, pp. 401- 407. http://dx.doi.org/10.1007/s11284-007-0393-9

9. C. Caratini and C. Tissot, "Paleogeographical Evolution of the Mahakam Delta in Kalimantan, Indonesia During the Quaternary and Late Pliocene," Review of Palaeobotany and Palynology, Vol. 55, No. 1-3, 1988, pp. 217-228. http://dx.doi.org/10.1016/0034-6667(88)90087-5

10. M. G. Sassi, A. J. F. Hoitink, B. de Brye, B. Vermeulen and E. Deleersnijder, "Tidal Impact on the Division of River Discharge over Distributary Channels in the Mahakam Delta," Ocean Dynamics, Vol. 61, No. 12, 2011, pp. 2211-2228.http://dx.doi.org/10.1007/s10236-011-0473-9

11. G. Lamorey and E. Jacobson, "Estimation of Semivariogram Parameters and Evaluation of the Effects of Data Sparsity," Mathematical Geology, Vol. 27, No. 3, 1995, pp. 327-358.http://dx.doi.org/10.1007/BF02084606

12. P. J. Curran, "The Semivariogram in Remote Sensing: An Introduction," Remote sensing of Environment, Vol. 24, No. 3, 1998, pp. 493-507. http://dx.doi.org/10.1016/0034-4257(88)90021-1

13. J. Yasrebi, M. Saffari, H. Fathi, N. Karimian, M. Moazallahi and R. Gazni, "Evaluation and Comparison of Ordinary Kriging and Inverse Distance Weighting Methods for Prediction of Spatial Variability of Some Soil Chemical Properties," Research Journal of Biological Sciences, Vol. 4, No. 1, 2009, pp. 93-102.

14. Govaerts and A. Vervoort, "Geostatistical Interpolation of Soil Properties in Boom Clay in Flanders," In: P.M. Atkinson, C.D. Lloyd, Eds., GeoENV VII—Geostatistics for Environmental Applications, Quantitative Geology and geostatistics, Vol. 16, Springer, Netherlands, 2010, pp. 219-230. http://dx.doi.org/10.1007/978-90-481-2322-3_20

15. K. Chang, "Introduction to Geographic Information System," 3rd edition, McGraw-Hill Publishing, New York, 2006, p. 432.

16. J. H. Zar, "Biostatistical Analysis," 4th edition, PrenticeHall, New Jersey, 1999, p. 469.

17. J. W. Osborne, "Improving Your Data Transformation: Applying the Box-Cox Transformation," Practical Assesment, Research & Evaluation, Vol. 15, No. 12, 2010, pp. 1-9. http://pareonline.net/genpare. asp?wh=0&abt=15

18. Ploner, "The Use of the Variogram Cloud in Geostatistical Modeling," Environmetrics, Vol. 10, No. 4, 1999, pp. 413-437.

19. A.L Jones, S.L. Kramer and P. Arduino, "Estimation of Uncertainty in Geotechnical Properties for Performancebased Earthquake Engineering, Pacific Earthquake Engineering Research Center, University of California, Barkeley, 2002, p. 23.http://peer.berkeley.edu/publications/ peer_reports/reports_2002/0216.pdf

20. X. Yang and T. Hodler, "Visual and Statistical Comparison of Surface Modelling Techniques for Point-based Environmental Data," Cartography and Geographic Information Science, Vol. 27, No. 2, 2000, pp. 165-75. http://dx.doi.org/10.1559/152304000783547911

21. J. E. A. Storms, R. M. Hoogendoorn, R. A. C. Dam, A. J. F. Hoitink and S. B. Kroonenberg, "Late-Holocene Evolution of the Mahakam Delta, East Kalimantan, Indonesia," Sedimentary Geology, Vol. 180, No. 3-4, 2005, pp. 149-166.http://dx.doi.org/10.1016/j.sedgeo.2005.08.003

22. N. G. Juma, "The Pedosphere and It's Dynamics. A System Approach to Soil Science. Volume 1. Intoduction to Soil Science and Soil Resources," Salman Production, Canada, 2001. http://www.pedosphere.ca/volume01/ pdf

23. McCauley, C. Jones and J. Jacobsen, "Soil and Water Management Module 1: Basic Soil Properties," Montana State University Extension Service, 2005.http://landresources.montana.edu/SWM/PDF/Final_ proof_SW1.pdf

24. L. D. Talley, "Salinity Patterns in the Ocean," In: M. C. MacCracken and J. S. Perry, Eds., Encyclopedia of Global Environmental Change. Volume 1: The Earth System: Physical and Chemical Dimensions of Global Environmental Change, John Wiley & Sons, 2002, pp. 629- 640.

25. D. Setyawan and Y. I Ulumuddin, 2012,"Species Diversity and Distribution of Bruguiera in Tambelan Islands, Natuna Sea, Indonesia," Proceedings of the Society for Indonesian Biodiversity International Conference, Society for Indonesian Biodiversity, Solo, 22-23 July 2011, Vol. 1. 2012, p. 290. http://biosains.mipa.uns.ac.id/P/index.htm

26. W. Giesen, S. Wulffraat, M. Zieren and L. Scholten, "Mangrove Guidebook for Southeast Asia," FAO and Wetlands International, 2006.

27. C. H. Chaineau, J. Mine and Suripno, "The Integration of Biodiversity Conservation with Oil and Gas Exploration in Sensitive Tropical Environments," Biodiversity Conservation, Vol. 19, No. 2, 2010, pp. 587-600. http://dx.doi.org/10.1007/s10531-009-9733-0

Chapter 5

REMOTE SENSING OF ENVIRONMENTAL CHANGE IN THE ANTIRIO DELTAIC FAN REGION, WESTERN GREECE

Emmanuel Vassilakis

Department of Dynamics, Tectonics and Applied Geology, Faculty of Geology & Geoenvironment, National & Kapodistrian University of Athens, Panepistimioupoli Zografou, 15784, Athens, Greece

ABSTRACT

In the westernmost region of the rapidly widening Corinth rift, Greece, extensive development of roads, bridges and other human infrastructure has caused continuous environmental change over the past twenty years. River networks, the land surface and the coastal environment, have been altered, especially in the areas corresponding to deltaic fans. In this paper we use earth observation systems that have captured these environmental changes, particularly medium (Landsat TM and ETM+) and high (Quickbird) resolution satellite images, to identify environmental changes between the periods 1992, 2000, 2002, and 2005. Six pseudo-color multi-temporal images in different spectral areas were created in order to detect changes to the terrestrial and coastal environment caused mainly by direct or indirect human impact. This methodology provided new data for quantifying significant alterations in the environment on different scales. In many cases this revealed their sequence during the time of observation.

INTRODUCTION

Over the last two decades the Greek government has proceeded with infrastructure development within the western part of Greece. Formerly there was no uninterrupted highway that connected mainland Greece in the north with the Peloponnesus in the south, and the Gulf of Corinth interrupted the land continuity. Until recently, the only direct modern road connecting these large parts of the country was on the easternmost side of the gulf where the natural strait is located. Transportation across the western part of the gulf consisted of small ferries completing the route between Rio (mainland Greece) and Antirio (Peloponnesus). This distance, of not longer than two kilometers, was covered by a ferry in about 15 minutes, not including the time for embarkation and disembarkation, but only during adequate weather conditions.

The Gulf of Corinth is one of the most rapidly spreading rifts in the world, generating large and disastrous earthquakes [1,2,3,4,5]. Thus, one of the main arguments against construction of a bridge across the gulf was the high seismic potential of the area, which is related to its rapid expansion, determined by GPS techniques to be an average of 18 ± 2 mm/yr in an approximately north-south direction across the gulf [6,7,8,9]. New construction techniques allowed for design of the Rio-Antirio Bridge, which was structurally supported by different fault blocks. Its design allowed the bridge to remain intact during future differential movements of the fault blocks [10]. The bridge was finally completed in 2004, providing a modern alternative highway along the western part of Greece. The anticipated increase in traffic along the highways leading to the bridge encouraged the local authorities to extend and widen the existing road network to the north-east and the west, and especially north of the bridge in the mainland sector. Accordingly, several construction sites were developed in the area surrounding the Antirio deltaic fan (Figure 1). Some of these are still ongoing as planned (quarries, highways, junctions, tunnels, *etc.*) and a high quality highway will be completed during the next decade, providing easier, safer and faster transportation in western Greece.

Figure 1: The study area on a natural color (3,2,1—R,G,B) reference image acquired by Landsat 5 on September 2, 1992. The Quickbird image used in this study (1,4,3—R,G,B) is shown as a partially overlapping layer. The white rectangle in the inset shows the location of the study area.

In addition to these sites supporting new highway construction, a new river dam was created to supply fresh water to the eastern part of mainland Greece from the mountainous and densely forested area of western Greece where precipitation is much higher. The dam was constructed on the Evinos River beginning in 1992 and was completed in June 2001. It was filled in October 2002. Water from this dam flows through an artificial tunnel to an older dam (on the Mornos River), before it reaches the network that supplies water to Athens and the surrounding cities. Therefore, in the valley downstream from the dam on the Evinos River, there has been a reduction in the average volume of water that reaches the coastal area just west of the new Rio-Antirio Bridge, along with a reduction in the sediment load carried by the river. Moreover, in the forested area between the Evinos valley and Antirio deltaic fan many wildfires—whether set intentionally or naturally occurring—have occurred over the last several decades. This has had serious consequences for the natural environment because most of the fire damage has not been repaired. Due to the high precipitation in this area, high rates of erosion have been recorded during wintertime, resulting in a large amount of transported material being deposited in a violent way along the Antirio deltaic coastal area [11,12].

This paper concerns the results of a first effort to document the surface changes of this tectonically and geomorphologically active region and to categorize them systematically, for further interpretation. The initial idea was to define whether the changes detected on the earth's surface are caused by the natural procedures of the evolving landscape or due to human interference, or both. The most efficient way to observe environmental alterations on such a wide scale, is the use of various remote sensing data, especially medium to high resolution and in the next section the methodology is described in detail.

DATA PROCESSING AND ANALYSIS

In order to detect environmental changes and the implications caused by the large construction sites along the Antirio fan and surrounding areas, several series of earth observation datasets were created. The satellite images used for this purpose were captured during several time periods beginning in 1992 and ending in 2005, and covering most of the area that was affected by the development of new infrastructure, as described in the introduction. The satellite image processing employed in this study includes atmospheric corrections, co-registration of images and creation of multi-temporal data constructed by creating new datasets for every spectral channel. The resulting false color composites reveal environmental changes not only where the construction sites are located but also in the coastal areas where erosion and deposition occur. The higher resolution images were used to increase the spatial resolution of the color composites, accomplished by merging high resolution images with lower resolution images [13,14] for better evaluation and quantification of the environmental alterations.

The first scene was acquired by the Thematic Mapper sensor (TM), placed on board the Landsat 5 satellite, on September 2, 1992, and consists of path 184 and row 33. Processing begins with geometric ortho-rectification in Universal Transverse Mercator (UTM) projection (zone 34N) by including, in the correction procedure, a highly detailed DEM with 25 meters pixel size [15]. The latter was produced by digitizing topographic map contours at a scale of 1:50,000. Subsequently, a dataset with 1123 columns and 577 rows was created; this covered only a part of the total scene and corresponds to the study area. The same procedure and dimensions were also applied to two other datasets, one derived from a Landsat 7-ETM+ (Enhanced Thematic Mapper) acquired on June 28, 2000 and another from a Landsat 7-ETM+ acquisition on August 13, 2005. All Landsat data were downloaded from U.S. Geological Survey (USGS) Earth Resources Observation and Science (EROS) Center. The last two datasets were ortho-rectified by using the image to image methodology and taking as reference dataset, the oldest one having the same DEM for altitude reference. A high resolution Quickbird 2 image acquired on June 12, 2002 and covering the central part of the study area was also used for larger scale observations with the highest possible accuracy and was imported in a geodatabase for further study along with the other data. These data were ortho-rectified, pan-sharpened and registered to the Landsat 5 ortho-corrected dataset, which was used primarily as the referencing image (Figure 1). The co-registration procedure of all datasets was successful since the RMS error was calculated less than 0.5 pixels either in X or Y geographic axis.

Effects of the atmosphere on remotely sensed data are not considered "errors" because they are part of the signal received by the sensing device. However, it is very important to remove atmospheric effects, especially on visible channels, as it has been noted that the atmospheric contribution to the radiance received by a satellite forms a much greater percentage of the radiance leaving the target area, than in the case of infrared [16]. Because this study is focused on detecting changes in a specific region through time, the main analytical tool relies on ground measurements made over time. Therefore it is very important to correct the radiance values recorded by the sensor for the effects of the atmosphere [17]. The technique that was applied on the Landsat series of datasets was based on the minimum digital number of the middle infrared band 7 for each dataset [18].

In general, when images contain areas of low reflectance, such as clear water bodies and deep shadows caused by extreme relief discontinuities, the corresponding pixels have values very close to zero, especially in the middle infrared band 7 of the Landsat TM and ETM+ sensors. Considering that a water body has virtually no reflection, even at shorter wavelengths such as band 4 or band 5, the digital number of a pixel representing deep sea water in band 7 should not be significantly larger than zero [19]. If it is larger than zero, then the non-zero digital number is due to atmospheric effects. This enables one to identify the effects of the atmosphere over large water bodies. The existence of the large water bodies in the Gulf of Patras west of the study area, and the Gulf of Corinth east of the study area, are thus very useful for calculating atmospheric corrections. After computing the statistics over each band of every dataset, the histogram minimum value for band 7 was calculated. This value was subtracted from the digital number arrays of the other bands of the same dataset and the result was a new array with digital numbers shifted towards the zero value, and presumably representing data with atmospheric corrections. The statistics for every single band were updated and the histograms were saved.

MULTI-TEMPORAL INTERPRETATION

The acquisition dates of the three Landsat images used for the multi-temporal interpretation were almost ideal for the purpose of this study because nearly all of them were captured during the summertime (late June to early September) when the local weather conditions are similar.

The multi-temporal analysis procedure begins with the compilation of six datasets, each one containing a single spectral channel from the three Landsat images. Each of these datasets contains three bands with the same Landsat spectral channel for each acquisition date. The thermal infrared channels of the Landsat 5 and Landsat 7 were not used because the spatial resolution varies for the TM and ETM+ sensors and the thermal reflectance cannot be considered identical throughout long time periods. Thus, the interpretation continues separately for each spectral channel in order to detect changes in the absorption and reflection spectra for specific bandwidths. Subsequently, six pseudochromatic images were produced by using the earliest data in the red channel (R-1992), the latest data in the blue channel (B-2005) and the intermediate data in the green channel (G-2000).

The resultant color composite images for every channel are presented in such a way as to locate the areas with high pixel values for each of the red, green or blue colors. In these cases, the earth's surface changed significantly from one acquisition date to the next. These observations proved to be quite difficult to obtain from images collected during the intermediate observation period (2000), as restoration works for the altered land coverage (river flow, forest fire, quarry operations *etc.*) could have taken place before the capturing date of the last satellite image (2005).

In detail, areas colored nearly pure red in the pseudo-chromatic images reveal areas that suffered significant change after September 2, 1992 and remained unchanged until August 13, 2005. The almost pure green areas represent changes on the earth's surface which happened in the time window between November 3, 1992 and June 28, 2000; with the provision that before and after these dates the land coverage should be more or less similar, or at least the spectral attitude of the area should be the same. If full and successful environmental restorations of an altered area have occurred, these areas would appear as green in each of the six datasets. Otherwise the restoration could not be characterized as completely successful, because the spectral attitude of the rehabilitated area is not the same before and after the environmental alteration. Finally, nearly pure blue areas represent change that happened after June 28, 2000 and before August 13, 2005.

CHANGE DETECTION RESULTS

In the pseudo-color image produced from band 1, the most spectacular change is represented by a blue linear feature connecting mainland Greece to the Peloponnesus; this is the Rio-Antirio cable bridge completed in 2004 and inaugurated just before the Athens Olympic Games (Figure 2). The blue color indicates that this bridge did not exist when the first two satellite images were acquired but did exist when the third satellite image was acquired. Thus it is clearly depicted on the 2005 Landsat image, which has been assigned to the blue channel. Its structural towers can be clearly seen in the southern part of the Quickbird image, which was captured before 2004 and apparently during construction (Figure 1). Additionally, the new road junctions on both sides of the gulf next to the bridge are also blue.

In the same image, just northeast of the town of Nafpaktos, a number of blue linear features indicate parts of the new highway detour designed to minimize traffic increase in the town itself. North of Nafpaktos, the lighter blue features with reddish parts are probably several road widenings with roadside banks constructed before 2005 (blue) or trenches constructed after 1992 (red). A new road constructed after 2000 in the forested area (21°43′E, 38°26′N) is also visible as a blue polyline feature (Figure 2).

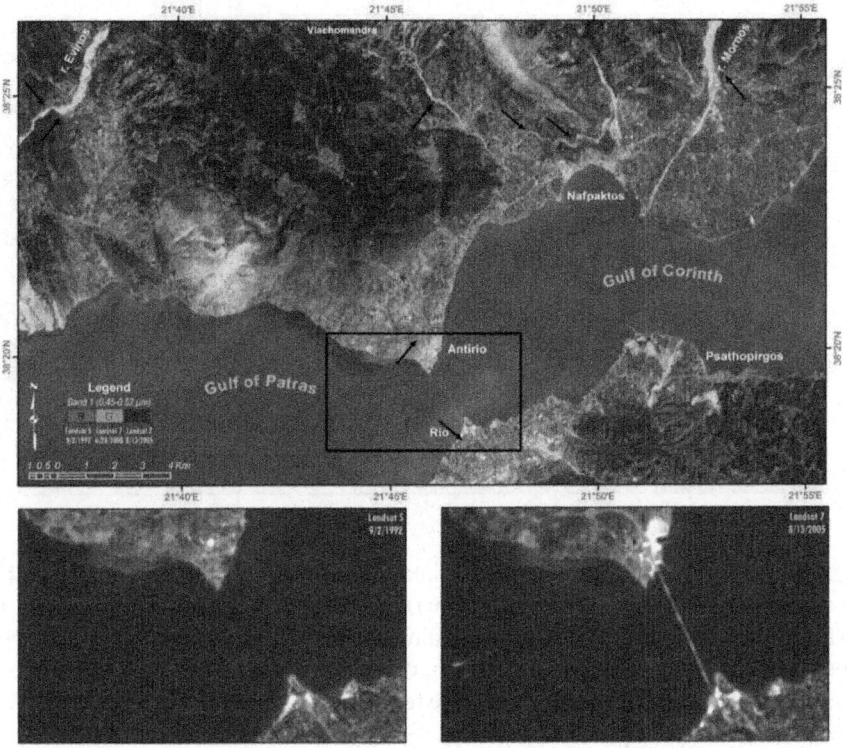

Figure 2: Multi-temporal pseudo-color image produced from layer stacking of Landsat satellite data (Band 1, 0.45 to 0.52 μm) with three different acquisition dates. The diachronic changes in onshore and offshore areas, largely due to infrastructure development after 2000, are clearly shown as blue linear features. The black arrows point to alterations of the surface described in the text. Below, two single band-1 enlargements of the black rectangle area show the changes caused by the bridge construction during the period from 1992 (left) to 2005 (right).

The flysch rocktype that makes up most of the bedrock in the study area has great potential for generating large-scale landslides. Frequent heavy rainfall in this area, combined with surface denudation from road construction and river erosion, cause major and minor land movements, which expose bare soil prior to rehabilitation and the renewal of vegetation (Figure 3) Such phenomena can be identified along the road that connects Nafpaktos with Vlachomandra, especially along the western side of the road where red pixels indicate landslides that happened shortly before 1992 and that were successfully restored before 2000. Larger-scale land movements seem to have occurred before 2005 on the eastern bank of the Mornos river (21°53′E, 38°25′N) and smaller scale changes along both banks of the Evinos river, especially when flysch lithologies were exposed along the river (21°36′E, 38°25′N).

Figure 3: Multi-temporal pseudo-color image produced from layer stacking of Landsat satellite data (Band 2, 0.52 to 0.60 μm) with three different acquisition dates. Land surface destruction due to mass landsliding along roads and river banks, as well as other changes in bare soil and vegetation, can be identified. The black arrows point to alterations of the surface described in the text.

A spectacular change in the land surface resulting from restoration of a burnt area can be observed over a large area in the north central portion of the study area (21°44'E, 38°24'N). In the oldest image (1992), this large area represents a region of bare soil in the middle of forest-type vegetation; such features normally correspond to tree destruction. In the multi-temporal pseudo-color images this area is colored reddish, whereas a smaller area in between the large ones is colored green (Figure 3 and Figure 4). In the high resolution Quickbird image (Figure 1) it is clear that in the same area there is a densely vegetated area with striped cultivation corresponding to vineyards. This interpretation is based on the reflectance contrast between bare soil and vegetation. Thus bare soil existed during 1992, a large part of which was cultivated before 2000 (red) and the vineyard expanded between 2000 and 2003 (green); as seen in the Quickbird image, the vegetation seems to be quite homogenous throughout this whole area.

Just west of this vineyard another major environmental alteration is indicated by the multi-temporal data analysis. A big quarry was developed sometime after 1992 but before 1996 when air photographs show its operation at a very early stage. This quarry is located on flysch, mainly consisting of shale and sandstone. Such materials are widely used in local housing settlements for decorative reasons. The quarry works have expanded, gradually causing a decrease in the forest vegetation that used to cover the surrounding area. Restoration of the quarry site must have begun between 2003 and 2005 because there is no blue in the false color image at this location, indicating that the reflectance was not that of bare soil when the latest Landsat image was acquired (Figure 4). There is also a smoothing of the relief as shown in the Quickbird image; this is not consistent with the exposure of the steep quarry terraces that are clearly visible in the 1996 high resolution air photographs.

Figure 4: Multi-temporal pseudo-color image produced from layer stacking of Landsat satellite data (Band 3, 0.63 to 0.69 μm) with three different acquisition dates. High contrast spots in the forested area reveal changes in reflectance due to human activity or forest fires. The time at which the reflectance changed can be estimated by the colors on the image, as described in the text (see black arrows).

For many years now, the reduction in water flow along rivers emptying into the Gulf of Corinth has caused coastal erosion and a regression of the shoreline. This could be related to construction of several dams along upstream portions of the rivers or perhaps to global climate change. Multi-temporal interpretation of Landsat images reveals that the northern coastlines of the two gulfs, on either side of Antirio, have suffered serious regression. In the near infrared datasets a discontinuous red stripe of a few pixels width is visible, beginning just east of Nafpaktos through the west of its waterfront, indicating that the beach which existed there in 1992—and possibly for many years before—has eroded (Figure 4 and Figure 5).

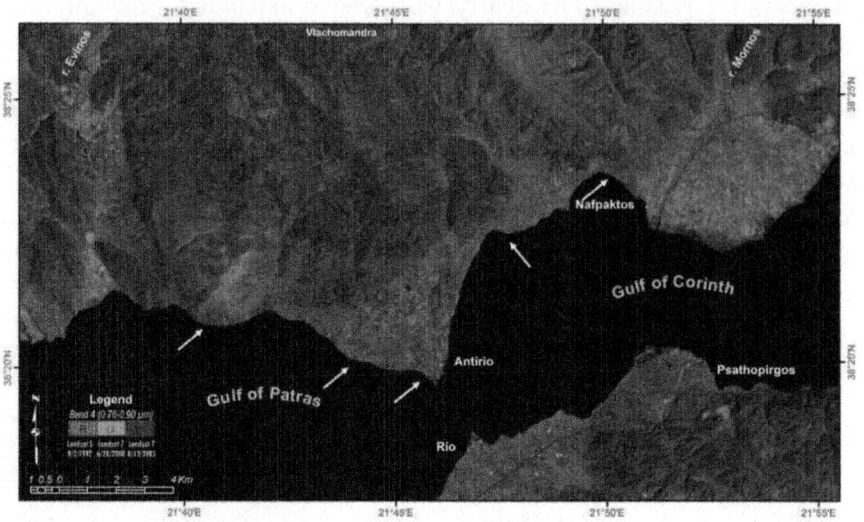

Figure 5: Multi-temporal pseudo-color image produced from layer stacking of Landsat satellite data (Band 4, 0.76 to 0.90 μm) with three different acquisition dates. The high absorbance of water creates a high reflectance contrast along the coastline so that shoreline regression is easily observed and quantified. The white arrows point to alterations of the coastal areas described in the text.

Along the eastern coast of the Antirio fan, no significant changes in the shoreline can be observed. A blue linear feature corresponds to a small pier constructed before 2005 but after 2000. In contrast, along the western part of the coastal deltaic area, an almost continuous red stripe corresponds to a regression of the 1992 shoreline, which in places has regressed by as much as 50 meters. Reddish pixels can also be observed along the western part of the coastline in the study area, where the main outcrops onshore consist of limestone, conglomerate, shale and sandstone, which make up parts of the flysch sequence. The steep topographic slopes present here, combined with the high frequency of planes of weakness in the rock (caused by bedding, joints, *etc.*) have the potential to produce rock fall into the sea; thus the red pixels along the coast line might represent missing land surfaces destroyed by rock fall, with rock fall probably occurring after 2000 and definitely after 1992. The epicenter of the large Aigion earthquake on June 15, 1995 (M=6.1) is located just 35 km eastward of this area, suggesting that rock fall could have been triggered directly by the earthquake or be related to post seismic surface deformation [20].

The Evinos dam, which is not located in the study area, has apparently altered the natural flow regime of the river. This has created temporary changes in the pattern of river incision, especially in the low relief segments of the valley at the northwestern end of the study area. In the multi-temporal pseudo-color images these changes are visible as linear features of different colors and widths; these linear features cross each other in a wide part of the river valley (Figure 6). These shifts in the river bed and incision pattern could have been the result of short-term meteorological changes, but because the three multi-temporal images were all acquired during the summer it is most likely that they are caused by controlled water release at the Evinos dam. Controlled water release at the dam would produce a significant change in the rate of discharge and sediment carrying capacity in downstream portions of the river, potentially causing erosion, landsliding and/or rock fall in the lithological formations that outcrop adjacent to the river valley, especially after sudden flooding.

Similar phenomena do not seem to have occurred along the Mornos river valley, perhaps because this dam has been in operation since 1981 and did not alter its operations during the period when the satellite images were acquired, or because a narrow artificial watercourse was constructed downstream of the dam in order to avoid floods at the river outlet in the Mornos deltaic fan area.

Figure 6: Multi-temporal pseudo-color image produced from layer stacking of Landsat satellite data (Band 5, 1.55 to 1.75 μm) with three different acquisition dates. The pattern of water flow in the rivers has shifted throughout the thirteen year observation period as indicated by crosscutting linear features of different colors. The black arrows point to alterations of the surface described in the text.

Because the lithology of the mountainous area consists mainly of flysch [21] and the climate is generally humid, the forest is extremely dense. During the summer whilst temperatures are high, forest fires occur quite frequently, resulting in large burnt areas with consequent environmental effects. By using the multi-temporal interpretation technique of Landsat data images, the areas affected by forest fires can by highlighted and one might be able to determine the extent of the burned area and the approximate time of the fire. In the case of Antirio, several burnt areas are apparent on the image in Figure 7.

A bluish area just east of the Evinos river valley corresponds to a burnt area that appeared between 2000 and 2005; there has been no effective rehabilitation. There are also many spotty areas in the forests that display a mixture of colors; these are in contrast with the densely vegetated areas that were burned and subsequently restored between the three capturing dates. For a more detailed study in terms of time, a more complete time series of images should be used in the future by applying the same methodology but with shorter periods of time between the capturing dates of the acquired remote sensing data.

Figure 7: Multi-temporal pseudo-color image produced from layer stacking of Landsat satellite data (Band 7, 2.08 to 2.35 μm) with three different acquisition dates. The reflectance in this band highlights the difference between bare soil and cultivated land, thus showing areas that have been denuded by forest fires. The various colors show surface changes between the three satellite acquisition dates. In particular, large areas of a reddish color appear across the mountainous area where previously burnt areas have been rehabilitated either naturally or artificially before the 2000 Landsat data was acquired. The black arrows point to alterations of the surface described in the text.

CONCLUSIONS

For the last decade, Greece has been making an ongoing investment in its infrastructure in western mainland Greece, including the building of roads, bridges and dams. The human interaction with the natural landscape evolution has a much higher impact than if the human factor was less significant. Almost all the landscape evolution observed for this area is directly or indirectly connected to major ongoing construction activities.

The environmental impact of these activities would not be apparent without the interpretation of remote sensing data such as described in this paper. The region of western Greece is nearly ideal for applying techniques involving multi-temporal remote sensing data because in a very short time period many environmental alterations have occurred due to natural and human activities. The multi-temporal image interpretation technique allows for the identification and, in many cases, the quantification of these environmental effects. The availability of a dense time series of remote sensing multispectral data provides an opportunity to identify changes on the earth's surface. In some cases, these may signal major environmental changes of a serious nature and indicate the need for preventative measures to avoid an environmental disaster.

This paper highlights the important role that analysis of multispectral satellite data can play in the identification of surface alterations related to human activity and natural processes. In some cases high resolution satellite images will also be needed to quantify the spatial extent of surface alterations, but multispectral data, at either medium or high spatial resolution, will remain the key in identifying the existence and timing of changes to the earth's surface environment.

ACKNOWLEDGEMENTS

The author would like to thank L. H. Royden for her constructive comments on an early version of this manuscript, as well as two anonymous reviewers whose suggestions and comments highly improved the text and the figure settings. Special thanks are due to D. J. Papanikolaou for encouraging the publication of this study and E. L. Lekkas for kindly offering the Quickbird raw data.

REFERENCES

1.   Jackson, J.A.; Gagnepain, J.; Houseman, G.; King, G.; Papadimitriou, P.; Soufleris, C.; Virieux, J. Seismicity, normal faulting, and the geomorphological development of the Gulf of Corinth (Greece) the Corinth earthquakes of February and March 1981. *Earth Planet. Sci. Lett.* 1982, *57*, 377–397.

2. Bernard, P.; Briole, P.; Meyer, B.; Lyon-Caen, H.; Gomez, J.-M.;
    Tiberi, C.; Berge, C.; Cattin, R.; Hatzfeld, D.; Lachet, C.; Lebrun, B.;
    Deschamps, A.; Courboulex, F.; Larroque, C.; Rigo, A.; Massonnet,
    D.; Papadimitriou, P.; Kassaras, J.; Diagourtas, D.; Makropoulos, K.;
    Veis, G.; Papazisi, E.; Mitsakaki, C.; Karakostas, V.; Papadimitriou, E.;
    Papanastassiou, D.; Chouliaras, M.; Stavrakakis, G. The M_s=6.2, June
    15, 1995 Aigion earthquake (Greece): evidence for low angle normal
    faulting in the Corinth rift. *J. Seismol.* 1997, *1*, 131–150.

3. Evangelidis, C.P.; Konstantinou, K.I.; Melis, N.S.; Charalambakis, M.;
    Stavrakakis, G.N. Waveform Relocation and Focal Mechanism Analysis
    of an Earthquake Swarm in Trichonis Lake, Western Greece. *Bull.
    Seismol. Soc. Amer.* 2008, *98*, 804–811.

4. Gaki-Papanastassiou, K.; Papanastassiou, D.; Maroukian, H.
    Geomorphic and archaeological-historical evidence for past earthquakes
    in Greece. *Ann. Geofis.* 1996, *39*, 589–601.

5. Papanikolaou, D.; Chronis, G.; Lykousis, V.; Pavlakis, P. Active Tectonics
    in the Rion-Antirion Strait, Western Greece. In Proceedings of 5th
    Meeting European Geological Societies, Dubrovnik, Croatia, October
    6–7, 1987; pp. 72–73.

6. Briole, P.; Rigo, A.; Lyon-Caen, H.; Ruegg, J.C.; Papazissi, K.;
    Mitsakaki, C.; Balodimou, A.; Veis, G.; Hatzfeld, D.; Deschamps, A.
    Active deformation of the Corinth rift, Greece: Results from repeated
    Global Positioning System surveys between 1990 and 1995. *J. Geophys.
    Res.* 2000, *105*, 25605–25625.

7. Avallone, A.; Briole, P.; Agatza-Balodimou, A.M.; Billiris, H.; Charade,
    O.; Mitsakaki, C.; Nercessian, A.; Papazissi, K.; Paradissis, D.; Veis, G.
    Analysis of eleven years of deformation measured by GPS in the Corinth
    Rift Laboratory area. *Comptes Rendus Geosciences* 2004, *336*, 301–311.

8. Bernard, P.; Lyon-Caen, H.; Briole, P.; Deschamps, A.; Boudin, F.;
    Makropoulos, K.; Papadimitriou, P.; Lemeille, F.; Patau, G.; Billiris, H.;
    Paradissis, D.; Papazissi, K.; Castarθde, H.; Charade, O.; Nercessian, A.;
    Avallone, A.; Pacchiani, F.; Zahradnik, J.; Sacks, S.; Linde, A. Seismicity,
    deformation and seismic hazard in the western rift of Corinth: New insights
    from the Corinth Rift Laboratory (CRL). *Tectonophysics* 2006, *426*, 7–30.

9. Cianetti, S.; Tinti, E.; Giunchi, C.; Cocco, M. Modelling deformation
    rates in the western Gulf of Corinth: rheological constraints. *Geophys. J.
    Int.* 2008, *174*, 749–757.

10. Parcharidis, I.; Foumelis, M.; Kourkouli, P.; Wegmuller, U. Persistent
    Scatterers InSAR to detect ground deformation over Rio-Antirio area

(Western Greece) for the period 1992–2000. *J. Appl. Geophys.* 2009, *68*, 348–355.

11. Zelilidis, A. The geometry of fan-deltas and related turbidites in narrow linear basins. *Geol. J.*2003, *38*, 31–46.

12. Bell, R.E.; McNeill, L.C.; Bull, J.M.; Henstock, T.J. Evolution of the offshore western Gulf of Corinth. *Geol. Soc. Amer. Bullet.* 2008, *120*, 156–178.

13. Welch, R.; Ehlers, W. Merging Multiresolution SPOT HRV and Landsat TM Data. *Photogramm. Eng. Remote Sensing* 1987, *53*, 301–303.

14. Rigol, J.; Chica-Olmo, M. Merging remote-sensing images for geological-environmental mapping: application to the Cabo de Gata-Nvjar Natural Park, Spain. *Environ. Geol.* 1998, *34*, 194–202.

15. Pouncey, R.; Swanson, K.; Hart, K. *ERDAS Field Guide*, 5th ed.; ERDAS Inc: Atlanta, GA, USA, 1999; p. 671.

16. Cracknell, A.; Hayes, L. *Introduction to Remote Sensing*; Taylor and Francis: London, UK, 1991; p. 293.

17. Mather, P. *Computer Processing of Remotely-Sensed Images: An Introduction*; Wiley & Sons: West Sussex, UK, 1994; p. 352.

18. Hadjimitsis, D.G.; Clayton, C.R.I.; Hope, V.S. An assessment of the effectiveness of atmospheric correction algorithms through the remote sensing of some reservoirs. *Int. J. Remote Sens.*2004, *25*, 3651–3674.

19. Lillesand, T.; Kiefer, R. *Remote Sensing and Image Interpretation*, 3th ed.; Wiley & Sons: New York, NY, USA, 1994; p. 750.

20. Parcharidis, I.; Metaxas, C.; Vassilakis, E. Earth observation data and geographical information system (GIS) techniques for earthquake risk assessment in the western Gulf of Corinth, Greece.*Canad. J. Remote Sens.* 2006, *32*, 223–227.

21. Jenkins, D. Structural development of Western Greece. *AAPG Bull.* 1972, *56*, 128–149.

Chapter 6

DETECTION AND MONITORING OF ACTIVE FAULTS IN URBAN ENVIRONMENTS: TIME SERIES INTERFEROMETRY ON THE CITIES OF PATRAS AND PYRGOS (PELOPONNESE, GREECE)

Issaak Parcharidis [1], Sotiris Kokkalas [2], Ioannis Fountoulis [3] and Michael Foumelis [4]

[1]Harokopio University of Athens, Department of Geography, El. Venizelou 70, 17671 Athens, Greece

[2]University of Patras, Department of Geology, Division of Physical Geology, Marine Geology and Geodynamics, 265 00 Patras, Greece

[3]National and Kappodistrian University of Athens, Faculty of Geology and Geoenvironment, Department of Dynamic Tectonic and Applied Geology, Panepistimioupolis Zografou, 157 84 Athens, Greece

[4]National and Kappodistrian University of Athens, Faculty of Geology and Geoenvironment, Department of Geophysics and Geothermics, Panepistimioupolis Zografou, 157 84 Athens, Greece

ABSTRACT

Monitoring of active faults in urban areas is of great importance, providing useful information to assess seismic hazards and risks. The present study concerns the monitoring of the potential ground deformation caused by the active tectonism in the cities of Patras and Pyrgos in Western Greece. A PS interferometric analysis technique was applied using a rich data–set of ERS–1 & 2 SLC images. The results of the interferometric analysis were compared with the tectonic maps of the two cities. Patras show clearer uplift–subsidence results due to the more distinct fault pattern and intense deformation compared to the Pyrgos area, where more diffused deformation is observed, with no significant displacements on the surface.

INTRODUCTION

The earthquake cycle of an active fault may include coseismic rupture and interseismic deformation. During the interseismic stage that usually ranges from a few hundreds to thousands of years, crustal tectonic strain may be silently accumulated. The strain is released during the interseismic period, especially along creeping active faults.

Understanding active tectonic processes and related energy release through monitoring of the transient deformation of strain accumulation process has become fundamental for several human activities. Additionally, local deformation type of a fault and the area near the fault may determine the extent of the seismic hazard as well. This allows taking into consideration measures and activities for seismic hazard mitigation.

Monitoring of active faults' interseismic behavior in urban areas is of great importance, as the local exposure (population, infrastructures etc.) increases the risks. Recently, interseismic crustal velocities and strains have been determined for a number of active areas, through repeated measurements using a Global Position System. In some cases the terrain is remote and the accessibility is difficult and thus the density of GPS measurements is relatively sparse, or in the case of urban environments, the operation of GPS receivers may be interrupted due to the frequent blockage of signals [1,2].

During the last two decades the SAR Differential Interferometric (DInSAR) technique based on radar satellite data has become a useful tool for ground deformation detection and monitoring [3,4]. Recent developments (since the end of the 90s) in DInSAR have demonstrated the potential to overcome some of the known limitations of repeat–pass interferometry. By examining interferometric phase from stable, point like targets, it is possible to monitor stability and cover an area that is normally characterized by low coherence. Additionally, millimetric target displacement along the line of sight (LOS) directions can be detected allowing the measurement of slow terrain motion [5]. Permanent or Persistent Scatterers Interferometry (PSI) is a technique used to calculate fine motions of individual ground and structure points over wide areas. These reflectors should remain stable (interferometric phase stability over time). Interferometric Point Target Analysis (IPTA) is a specific method of PSI to exploit temporal and spatial characteristics of interferometric signatures collected from point targets to map scatterer deformation history [6]. Monitoring faults in urban environments using different interferormetric techniques (repeat–pass, stacking and PS) has been widely used in seismically active areas of the globe [1,7,8,9,10,11,12,13,14].

During 1993 catastrophic earthquakes affected northwestern Peloponnese in the southern part of Greece. They caused serious damages to the greater area of two densely populated cities, Patras and Pyrgos. These are the capital cities of the neighboring prefectures of Achaia and Ilia. This study concerns the use of the PS interferometric analysis of ERS–1 and 2 satellite data, over the cities of Pyrgos and Patras in order to monitor and reveal the spatial distribution of creep along active faults.

TECTONIC AND GEOLOGICAL FRAMEWORK—ACTIVE TECTONICS

Western Greece is an area that characterized by high seismicity and a fairly complex three-dimensional setting, with along strike changes in the progress of subduction zone. It comprises continent–continent collision in the north and ocean–continent subduction in the south [15,16,17,18,19]. The change between continent–continent and ocean–continent subduction occurs at the Kephalonia Transform Fault (KTF) [20], which is characterized by a dextral strike–slip sense of shear. GPS data from western Greece show a small amount of motion north of the KTF and southwest-directed rapid motion of the overriding plate south of it that reaches rates of 30–35 mm/yr in the western parts of the Peloponnese along with dextral kinematics of the KTF [21,22].

The NW Peloponnese about 70–80 km east of the present day NNW–SSE trending Hellenic Trench (Figure 1) has a complex geological history of tectonics and erosion. However, the active deformation in western Greece could be described in spatio–temporal continuation with the foreland–propagated fold and thrust belt of the External Hellenides, which can be followed along the Hellenic arc [15,19]. During the Eocene, the Peloponnese was characterized by an Alpine collisional history, which led to assemblage of intra-Tethyan continental fragments (e.g., Apulia and Pelagonian microcontinents) and the formation of the Hellenic mountain range [23]. Within this context, Mesozoic early Cenozoic carbonate rocks, originally deposited on a series of platforms (Pre-Apulian and Gavrovo isopic zones) and basins (Ionian and Pindos isopic zones), were telescoped by a system of N–S striking and east-dipping thrust faults that propagated upward and westward into overlying flysch deposits [24,25,26]. From the late Miocene and onwards thrust processes progressively shifted westward from the Peloponnese area to the Ionian Sea [15,27]. Compressional structures of Upper Miocene to Quaternary age have been recognized in Ionian Islands from field based studies, as well as in deep seismic profiles [15,28].

An extensional stress field, in turn, has prevailed in the northwestern Peloponnese from the early Pliocene up to present day generating three major sets of active normal faults with NE–SW, WNW–ESE and ENE–WSW trends, respectively [27,29,30,31,32]. The complexity of the structure in the westernmost end of the Hellenic subduction is imprinted on the great variety of earthquake focal mechanisms.

Patras

The Patras graben is a 40 km long and 15 km wide fault zone which extends from Cape Araxos in the west to the village of Hellinikon in the east. It comprises thick fluvio–deltaic conglomerates, and floodplain or shallow lacustrine sands and clays of Pleistocene age. Towards its eastern part these deposits pass gradually into a coarse conglomerate sequence that was formed as debris flow facies in a fan environment [27].

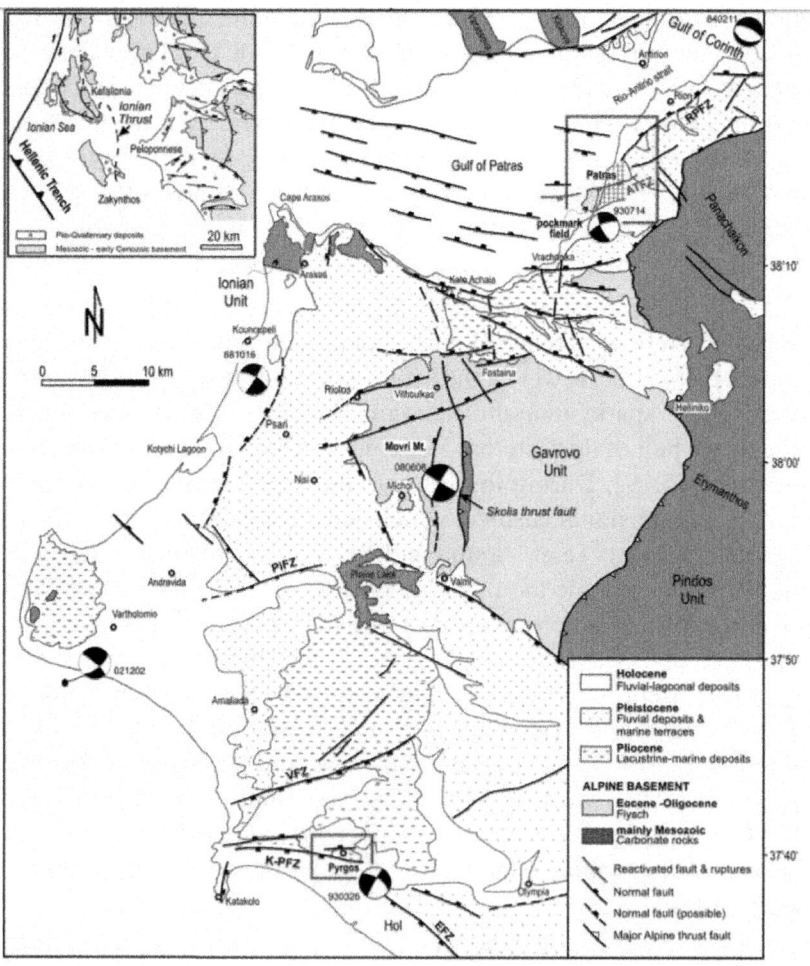

Figure 1: Tectonic setting of the broader study area (ATFZ: Ag. Triada Fault Zone, EFZ: Epitalio fault Zone, RPFZ: Rio–Patra fault Zone, K–PFZ: Katakolo– Pyrgos

Fault Zone, PiFZ: Pinios Fault Zone, VFZ: Vounargo Fault Zone, blue frames indicate the study areas, map modified from Koukouvelas *et al.* [50].

Holocene sediments, with a thickness of 40 m, are arranged in a series of tilted fault blocks bounded by WNW–trending normal faults that accommodate the extension in the area [33]. Generally, WNW trending faults are clearly segmented along strike and most are characterized by variable displacement. Such differences are accommodated by NNE–oriented transfer faults, which permit abrupt changes in depositional conditions along the WNW–trending downthrown blocks and bound small subbasins at the southern margin of the basin.

On the onshore part of the graben, a major WNW–trending fault scarp is formed on the southern part of the graben along the pre-Neogene margin. ENE–directed faults are also present and displace fluviatile terraces. Holocene net vertical slip on the central parts of Patras basin is about 3–5 mm/yr, while on its northern and southern margin is about 1.2 mm/yr [34]. Rifting in the Patras area may have initiated some 1.3–1.5 Ma ago [35].

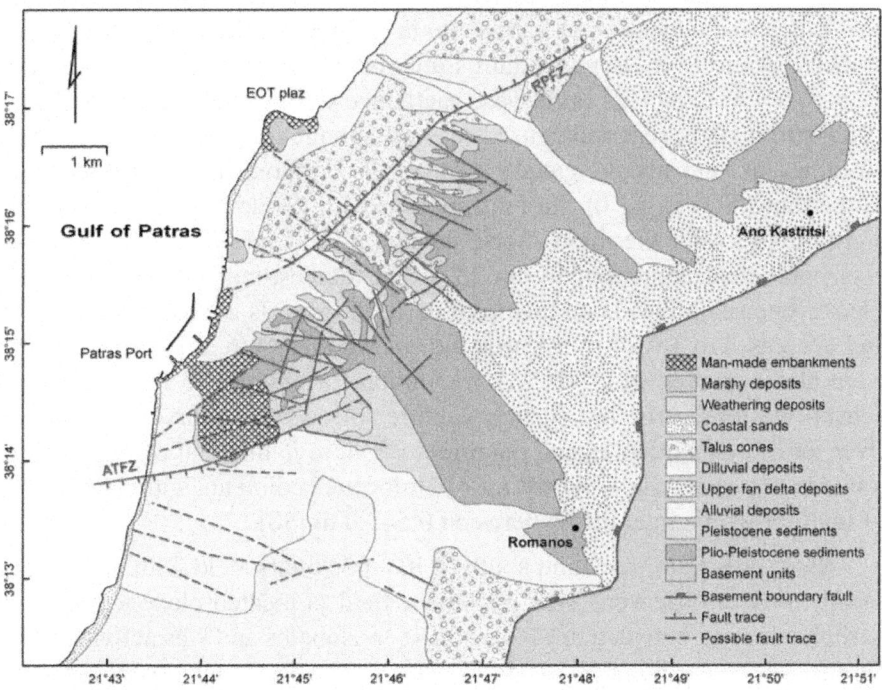

Figure 2: Geological map of Patras City (ATFZ: Ag. Triada Fault Zone, RPFZ: Rio–Patra fault Zone), map modified from Doutsos *et al.* and Koukis *et al.* [27,36].

Further to the east, the Rio graben is developed from the village of Vrahneika to the city of Patras and continues further east through Rio to Nafpaktos town. The Rio graben displays two distinct sedimentary cycles: (a) thick silts and clays deposited during Upper Pliocene into a lacustrine to shallow marine environment and (b) coarse fluvial deposits and alluvial fans of Pleistocene age. Pleistocene Rion series are found uplifted up to 500–600 m high in the hills of Ano Kastritsi. A fairly straight NE–trending fault slope limits this hilly morphology from the narrow Patras littoral plain. Major structural trends in this area vary between NE and ENE (55°–70°). Rio graben acted as a transfer zone between the differently extending Patras and Corinth grabens by reactivation of preexisting NE–SW trending faults in the Pliocene [27].

The city of Patras is founded mainly on Quaternary deposits and Plio–Pleistocene sediments (Figure 2) with a thickness exceeding 300 m, based on borehole data [36]. The fault trace map of the broader residential area of the city, based on fieldwork and air photo interpretation [36] shows a NE–SW main fault trend at the northern part of the city and a more prominent WNW–ESE trend in the southern part. This complex fault interplay in the area is due to the location of Patras city in the junction between Patras and Rio grabens. Some ENE–trending faults are also present, such as the Ag. Triada fault zone (ATFZ). This fault was reactivated during the August 31 (Ms 4.8) 1989 earthquake event and caused serious damage to new multistorey and old buildings, in a narrow elongated zone about 1.5 km long and 50 m wide parallel to the fault. Surveying of the fault motion with geodetic methods for almost eight months after the main shock showed a total subsidence of 25 mm and horizontal displacement of 14 mm [37]. The total estimated throw on the Plio–Pleistocene sediments is on the order of 40 m towards the eastern part of fault and decreases to 15–20 m towards the coastal areas [36,37]. Surface ruptures were also observed in a N70° E orientation for almost 1.5 km. Towards the coastal western parts, the surface rupture follows the course of Diakoniaris river for more than 500 m and continues offshore in the Gulf of Patras dipping towards the south, showing tilting of Holocene sediments and clear evidence of faulting with a total throw between 0.5–5.0 m [38].

Along this fault trace and south of it, a pockmark field with craters of gas expulsion was observed. This pockmark field is located close to Patras new harbor, confined between the 10 m and 45 m isobaths and was activated during another strong earthquake (Ms 5.4) on 1993. The earthquake focal mechanism [39] shows a N058° trending nodal plane. If this nodal plane is the active one, the slip on this plane is dextral with a dip–slip component. Anomalously temperature increase and gas expulsion along an ENE–pockmark string of craters recorded prior the earthquake activity suggesting that this fault system is active and plays an important role in fluid circulation [40].

Pyrgos

The city of Pyrgos is located about 70–80 km east of the NNW–SSE trending Hellenic Trench. Neotectonic structures in the area are not oriented parallel to the arc in the NNW–SSE direction [41]. Instead, there are several WNW–ESE, E–W, and NE–SW trending normal faults forming the margins of the post-Alpine basins developed on top of the thrust sheets of the Hellenides, with alternating horst and graben structures [32,42,43,44].

The broader Pyrgos–Olympia area corresponds to a large 1st order graben structure bounded to the north by the Erymanthos horst, to the east by the Tropea horst and to the south by the Lapithas Mt. horst all of which are built of alpine age sediments. In the Pyrgos–Olympia graben there are more than 3 km of Plio–Quaternary sediments, including some diapiric structures related to the existence of Upper Miocene evaporites at depth, detected from geophysical prospecting and drilling [45]. Continental and lacustrine Pliocene–Lower Pleistocene deposits occur mainly at the eastern part of the Olympia basin whereas marine sediments occurring along the coastal zone of the gulf are mainly of Lower Pleistocene age. The neotectonic evolution of the Pyrgos graben was not the same in all its extent [29]. It was differentiated due to the creation and evolution of smaller size tectonic blocks (2nd, 3rd order macrostructures). One of these smaller order structures is the Pyrgos horst. A great number of faults have been located and studied during neotectonic surveys; some of these were active in previous neotectonic periods (e.g., Late Miocene, Pliocene, Early Pleistocene), whereas others are active Holocene structures [46]. The E–W trending active faults in the Pyrgos area within the Pyrgos–Olympia basin have been analysed after the 1993 destructive earthquake event [30,47].

The outcrops in the broader Pyrgos area (Figure 3) consist of: (i) Alluvial deposits, which crop out in the flat area of the broader Pyrgos area and overlay unconformably the older formations, while their thickness does not exceed 12 m; (ii) Erymanthos Formation, which outcrops over a limited area and consists of red to brown red clays and yellow brown sandy clays of Pleistocene age. It overlies unconformably the Vounargo formation and its thickness varies from 2 m to 8 m; (iii) Vounargo Formation, comprises continuous intercalation of clays, silts, sandstones, sands and marls of Plio–Pleistocene age, with a total thickness up to 600 m [29,46]. The deposits that occur in the area of Pyrgos have undergone neotectonic deformation and are crossed by a number of faults of E–W mean strike that form part of the Katakolo–Pyrgos fault zone, which was responsible for the earthquakes of 26 March 1993 [29].

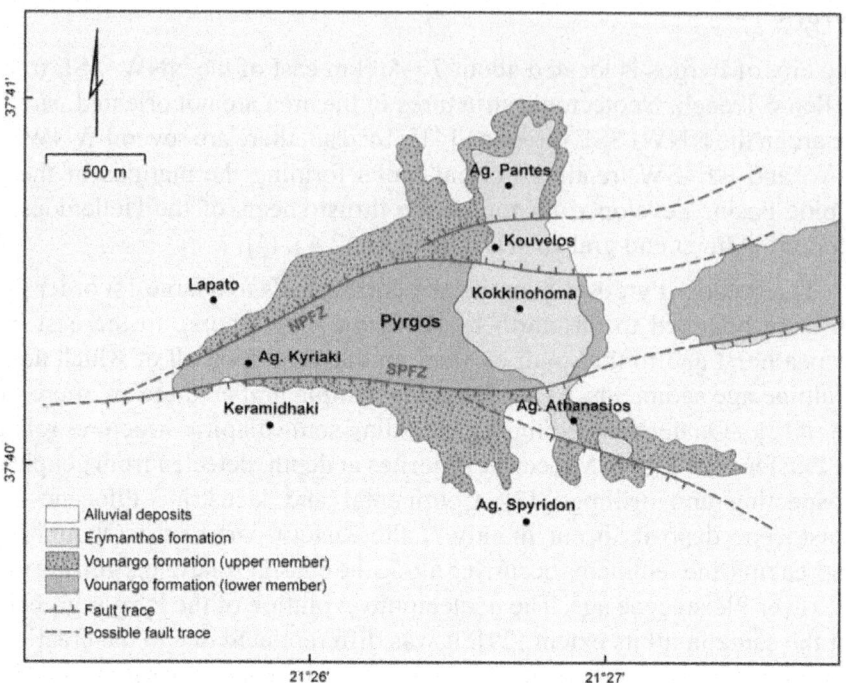

Figure 3: Geological map of Pyrgos City (NPFZ: North Pyrgos Fault Zone; SPFZ: South Pyrgos Fault Zone).

More specifically, the elongated outcrop of the lower member terminates at the South Pyrgos fault, south of which the upper member of the Vounargo formations occurs. It is a normal fault, accompanied by a morphological discontinuity (to the south of the city main square); its throw is at least 50 m (estimated from morphotectonic features) and eastwards it branches into two faults.

To the north there is an identical setting with the North Pyrgos fault (av. strike NE–SW), which also branches into two minor faults to the east. The North Pyrgos fault juxtaposes the outcrops of upper members, while it also crosses some outcrops of Erymanthos formation. Its throw is smaller, (20–30 m) and all along it, seismic fractures caused by the shock of 26 March 1993, were recognized.

The main shock as well as the aftershocks caused extensive damage in Pyrgos and the surrounding area. Several secondary geological effects such as liquefaction, sandblows, landslides and fractures, were observed.

SAR INTERFEROMETRIC PROCESSING AND ANASLYSIS

Interferometric processing for the selected study areas was performed using 42 ERS–1/–2 SAR scenes acquired between 1992 and 2000 along descending track 279 (Table 1) for the Patras case and 39 ERS–1/–2 SAR scenes along descending track 279 for Pyrgos (Table 2).

Table 1: ERS–1 & 2 scenes (Track = 279, Frame = 2835) that have been used (in bold the master reference image)

Count	Master (date)	Satellite	Slave (date)	Orbit	Bp (meters)	dT (days)
1	03/06/1995	ERS–1	12/11/1992	6937	73.9	−933
2	03/06/1995	ERS–1	10/06/1993	9943	−477.4	−723
3	03/06/1995	ERS–1	19/08/1993	10945	−323.7	−653
4	03/06/1995	ERS–1	28/10/1993	11947	569.7	−583
5	03/06/1995	ERS–1	25/03/1995	19305	−1169.9	−70
6	03/06/1995	ERS–1	29/04/1995	19806	−518.5	−35
7	**03/06/1995**	**ERS–1**	**03/06/1995**	**20307**	**0**	**0**
8	03/06/1995	ERS–1	08/07/1995	20808	−583.4	35
9	03/06/1995	ERS–2	13/08/1995	1636	82.9	71
10	03/06/1995	ERS–2	17/09/1995	2137	−360.1	106
11	03/06/1995	ERS–1	21/10/1995	22311	761.3	140
12	03/06/1995	ERS–2	31/12/1995	3640	186.2	211
13	03/06/1995	ERS–2	19/05/1996	5644	71.0	351
14	03/06/1995	ERS–2	23/06/1996	6145	−83.8	386
15	03/06/1995	ERS–2	01/09/1996	7147	−514.6	456
16	03/06/1995	ERS–2	06/10/1996	7648	−392.7	491
17	03/06/1995	ERS–2	10/11/1996	8149	1064.9	526
18	03/06/1995	ERS–2	15/12/1996	8650	−309.2	561
19	03/06/1995	ERS–2	19/01/1997	9151	−28.7	596
20	03/06/1995	ERS–2	23/02/1997	9652	−182.3	631
21	03/06/1995	ERS–2	04/05/1997	10654	−360.3	701
22	03/06/1995	ERS–2	08/06/1997	11155	−166.9	736
23	03/06/1995	ERS–2	13/07/1997	11656	−124.9	771
24	03/06/1995	ERS–2	17/03/1997	12157	111.5	806
25	03/06/1995	ERS–2	21/09/1997	12658	−263.2	841
26	03/06/1995	ERS–2	30/11/1997	13660	177.5	911
27	03/06/1995	ERS–2	04/01/1998	14161	116.4	946
28	03/06/1995	ERS–2	19/04/1998	15664	193.1	1051
29	03/06/1995	ERS–2	24/05/1998	16165	−160.7	1086
30	03/06/1995	ERS–2	28/06/1998	16666	−834.9	1121
31	03/06/1995	ERS–2	02/08/1998	17167	56.7	1156
32	03/06/1995	ERS–2	06/09/1998	17668	47.0	1191
33	03/06/1995	ERS–2	28/02/1999	20173	302.9	1366
34	03/06/1995	ERS–2	13/06/1999	21676	−559.8	1471
35	03/06/1995	ERS–2	18/07/1999	22177	455.9	1506
36	03/06/1995	ERS–2	22/08/1999	22678	1012,1	1541
37	03/06/1995	ERS–2	26/09/1999	23179	448.6	1576
38	03/06/1995	ERS–2	31/10/1999	23680	357.3	1611
39	03/06/1995	ERS–2	05/12/1999	24181	−121.9	1646
40	03/06/1995	ERS–2	09/01/2000	24682	−140.1	1681
41	03/06/1995	ERS–2	23/04/2000	26185	953.7	1786
42	03/06/1995	ERS–2	28/05/2000	26686	806.6	1821

Table 2: ERS–1 & 2 scenes (Track = 279, Frame = 2846) that have been used (in bold the master reference image)

Count	Master (date)	Satellite	Slave (date)	Orbit	Bp (meters)	dT (days)
1	17/8/1997	ERS–1	12/11/1992	6937	−44.7	−1739
2	17/8/1997	ERS–1	10/6/1993	9943	−581.5	−1529
3	17/8/1997	ERS–1	19/8/1993	10945	−432.0	−1459
4	17/8/1997	ERS–1	28/10/1993	11947	455.7	−1389
5	17/8/1997	ERS–1	25/3/1995	19305	−1274.8	−876
6	17/8/1997	ERS–1	29/4/1995	19806	−624.8	−841
7	17/8/1997	ERS–1	3/6/1995	20307	−107.5	−806
8	17/8/1997	ERS–2	13/8/1995	1636	−28.5	−735
9	17/8/1997	ERS–1	21/10/1995	22311	653.7	−666
10	17/8/1997	ERS–1	18/5/1996	25317	39.4	−456
11	17/8/1997	ERS–2	23/6/1996	6145	−196.6	−420
12	17/8/1997	ERS–2	1/9/1996	7147	−622.8	−350
13	17/8/1997	ERS–2	10/11/1996	8149	954.8	−280
14	17/8/1997	ERS–2	15/12/1996	8650	−415.3	−245
15	17/8/1997	ERS–2	19/1/1997	9151	−139.1	−210
16	17/8/1997	ERS–2	23/2/1997	9652	−288.6	−175
17	17/8/1997	ERS–2	4/5/1997	10654	−466.7	−105
18	17/8/1997	ERS–2	8/6/1997	11155	−276.1	−70
19	17/8/1997	ERS–2	13/7/1997	11656	−235.9	−35
20	**17/8/1997**	**ERS–2**	**17/8/1997**	**12157**	**0**	**0**
21	17/8/1997	ERS–2	21/9/1997	12658	−370.4	35
22	17/8/1997	ERS–2	30/11/1997	13660	63.3	105
23	17/8/1997	ERS–2	4/1/1998	14161	9.9	140
24	17/8/1997	ERS–2	19/4/1998	15664	78.8	245
25	17/8/1997	ERS–2	24/5/1998	16165	−271.3	280
26	17/8/1997	ERS–2	28/6/1998	16666	−941.3	315
27	17/8/1997	ERS–2	2/8/1998	17167	−47.8	350
28	17/8/1997	ERS–2	6/9/1998	17668	−60.7	385
29	17/8/1997	ERS–2	28/2/1999	20173	186.5	560
30	17/8/1997	ERS–2	13/6/1999	21676	−674.8	665
31	17/8/1997	ERS–2	18/7/1999	22177	341.6	700
32	17/8/1997	ERS–2	22/8/1999	22678	898.1	735
33	17/8/1997	ERS–2	26/9/1999	23179	345.9	770
34	17/8/1997	ERS–2	31/10/1999	23680	247.4	805
35	17/8/1997	ERS–2	5/12/1999	24181	−236.6	840
36	17/8/1997	ERS–2	9/1/2000	24682	−247.4	875
37	17/8/1997	ERS–2	23/4/2000	26185	838.9	980
38	17/8/1997	ERS–2	19/11/2000	29191	839.0	1190
39	17/8/1997	ERS–2	24/12/2000	29692	−567.1	1225

Initial estimates of the interferometric baselines were calculated from available precise orbit state vectors from Delft Institute (NL) for Earth–Oriented Space Research (DEOS) [48]. The topographic phase was simulated based on SRTM V3 DEM of approximate spatial resolution of 90 m, resampled to 40 m to fit the SAR data resolution.

Time-series analysis was based on the IPTA processing scheme [6]. Temporal and spatial characteristics of interferometric signatures collected from point targets are exploited to accurately map average ground deformation rates and deformation histories.

The selection of the reference points is regarded to be the most critical part of the IPTA, as final deformation rates and histories are greatly affected by this decision. Some criteria for the selection of the reference point are dictated by the applied method, such as the high quality of the point in terms of phase stability overtime. Others are related to the regional tectonic setting of the area and the related pattern of deformation which needs to be extracted.

Interferometric point target analysis

Starting from a stack of coregistered Single Look Complex (SLC) images, the selection of the reference scene was based primarily on the baseline minimization criteria. In addition, the selection of a reference scene acquired near the temporal average of the available SAR acquisitions is also of interest.

The first step of the analysis involves the identification of candidate point targets for which the time-series analysis will be performed. In this case two different approaches were applied. The first approach is based on the spectral properties of each individual SLC image. This is done by identifying point targets of low spectral phase diversity. The second approach involves the identification of candidate point targets based on low intensity variability, since by definition point targets do not show speckle behavior as simple coherent scatterer dominates the echo.

The analysis of the differential interferometric phases in the temporal direction is an important element of an interferometric point target analysis. Point data stack of differential interferograms was generated and analyzed by means of phase regression analysis in the temporal domain using two dimensional bilinear regression model.

Two-dimensional regression analysis is done with the dimensions corresponding to the perpendicular baseline of the interferometric pairs and to the time difference between the two SLC of the interferometric pairs according to:

$$a0 + a1 \times bperp\,[i] + a2 \times delta\_t\,[i]$$

where a0: phase offset, a1: slope in baseline dimension (can be converted into point height correction), a2: slope in time dimension (can be converted into linear deformation rate), bperp: perpendicular baseline component of interferograms, and delta_t: time interval of interferograms.

The model examines linear dependence of the topographic phase on the perpendicular baseline component as well as linear phase dependence with time, solving respectively for both height correction and constant deformation rate of the point target relative to the reference. The regression analysis of the entire stack of observations was first conducted using multiple patches, within each patch one reference is determined, and then using the selected single reference point as a global reference. This procedure was followed in order to minimize the effect of distance between the two pairs of phase components, as the atmospheric distortion, baseline error (residual orbital phase trends) and higher relative deformation rates result in higher deviations of the individual points from the regression plain. The quality of the preliminary candidate points were then carefully evaluated based on the estimated standard deviation of the differential interferometric phase from the 2-D regression model.

Points with a phase standard deviation larger than the indicated threshold (in this case 1.0) were rejected, significantly reducing the number of scatterers. The majority of the rejected points were located over mountainous areas. A total number of 6,613 and 1,829 targets for Patras and Pyrgos respectively were detected.

The general phase model for IPTA that used is the same as the conventional interferometry. The unwrapped interferometric phase Φunw is expressed as the sum of topographic phase Φtopo, deformation phase Φdef, path delay Φatm (atmospheric phase) and the phase noise Φnoise:

$$\Phi unw = \Phi topo + \Phi def + \Phi atm + \Phi noise$$

Phase terms related to the atmosphere, nonlinear deformation, baseline errors and noise can be discriminated within the residual phases based on their differing spatial and temporal dependencies [6].

Unwrapped phases calculated from the regression analysis described above and the corresponding topographic phases were then used in a least-squares approach for baseline refinement. The analysis was limited over areas exhibiting no deformation as dictated by the linear deformation estimates. Introducing the refined baselines and taking into consideration the early estimated height corrections and linear deformation rates, the interferometric phase model was updated in a second iteration.

Additional processing includes temporal and spatial filtering of newly estimated residual phases to compensate for atmosphere and noise. Atmospheric screen was attributed to large scale nonlinear residuals and subtracted from the model by applying low-pass spatial filtering on the residual phases. Phase noise was treated by spatially filtering of phases around the reference, assuming stability of the area considered.

Further iteration applying the additional corrections results in the final regression model. Results consist of point heights, linear deformation rates, atmospheric phase, refined baselines, quality information (temporal coherence) and nonlinear deformation histories for each point. It is important to mention that the final deformation model, as a consequence of the assumptions made during the estimation of atmospheric phase contribution to the signal, includes nonlinear components of only local scale phase variations.

RESULTS

After transformation of the interferometric results from range–Doppler coordinates into map geometry (geographic coordinates), point targets were imported in a GIS environment and plotted on a panchromatic Landsat–7 ETM+ image and in the Google Earth environment for point target identification (Figure 4 and Figure 5).

Here we should note that, although the reference points are considered stable there are no absolute stable points in the area. In this way, negative velocities do not necessarily represent subsidence, but possibly slower uplift according to the reference point. Thus, when we use the terms subsidence and uplift we refer to relative values with regard to the picked reference point.

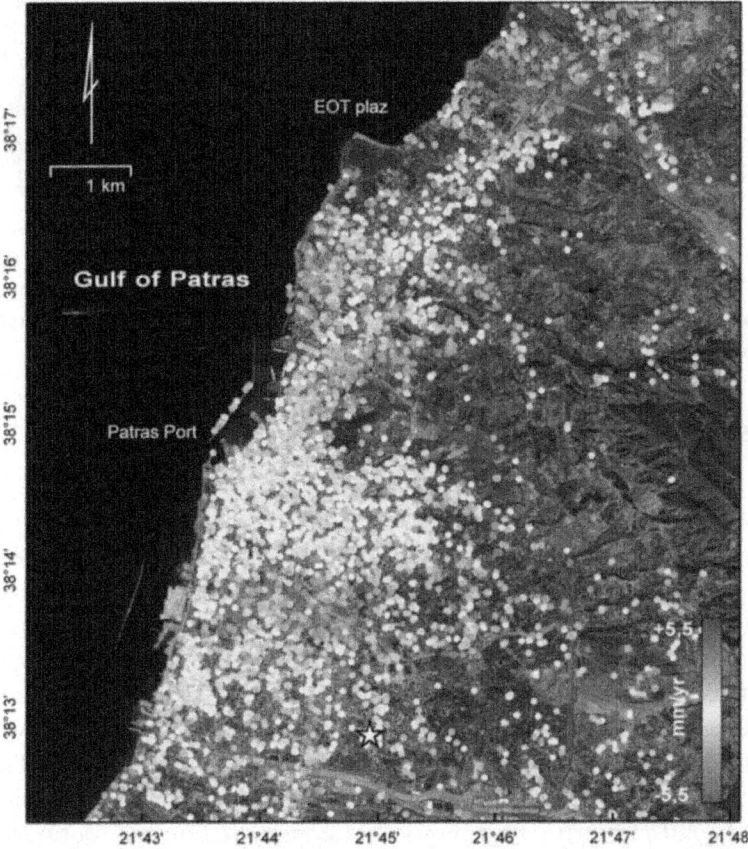

Figure 4: Linear component of ground deformation obtained by IPTA over Patras city. Point targets are plotted on a Panchromatic Landsat–7 ETM+ image. The star on the image refers to the location of the reference point.

Figure 5: Linear component of ground deformation obtained by IPTA over Pyrgos city. Point targets plotted on a QuickBird Pan–Sharpened image. The star on the image refers to the location of the reference point.

Patras

From Figure 6 it is indicated that generally the relative vertical velocities toward and away from the satellite range in the LOS, vary between maximum values of +5 mm/yr and −5 mm/yr, respectively. A remarkable aspect is that there is contrasting subsidence and uplift of PS points along discrete and specific zones, such as the Ag. Triada fault zone in the south of the city and along a W–E trending lineament, north of the Patras port, that doesn't relate with any visible mapped fault trace (Figure 7). Between these two zones only uplift even with low rates is observed. This area represents in a way the relay zone and the hanginwall block of both Ag. Triada and Rio–Patras fault zone, consisting of Plio–Pleistocene sediments. Maximum uplift velocities on the order of +4.0 mm/yr to +5.5 mm/yr are constricted on the footwall block of Pio–Patras fault, while in the relay zone between that fault zone and Ag. Triada fault lower values between +2.5 mm/yr and +4.0 mm/yr are observed. Similar values are also calculated towards west close to the basin bounding Kastritsi fault.

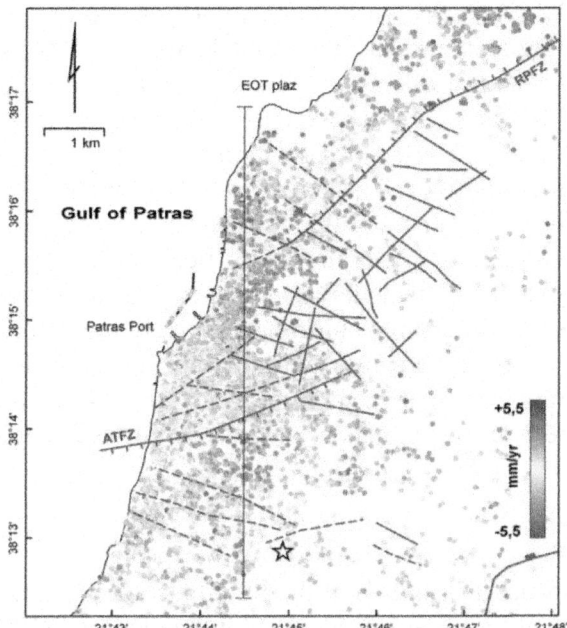

Figure 6: Point targets plotted over the fault map for Patras area (ATFZ: Ag. Triada Fault Zone), solid line corresponds to the section of figure 7, fault pattern modified from Koukis *et al.* [36].

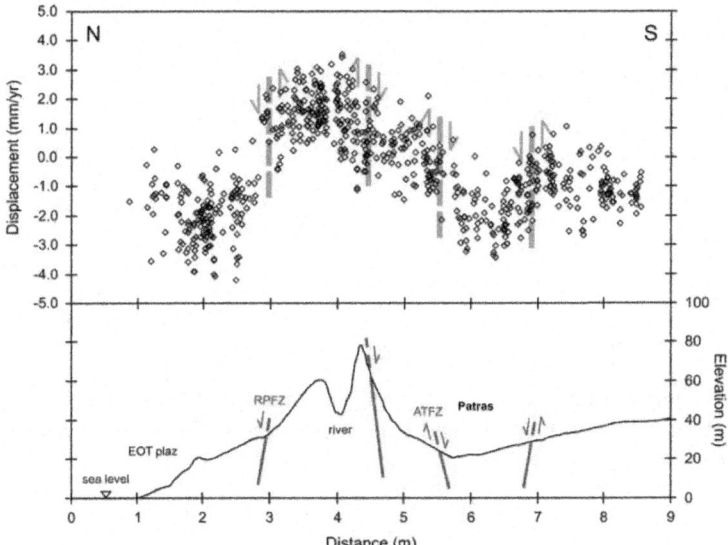

Figure 7: Spatial profile showing the displacement field as observed by IPTA and plotting targets within 150 meters in both sides of the profile (upper), and the related topography (down), red lines correspond to the faults.

Accordingly, maximum subsidence values on the order of −4.5 mm/yr to −5.5 mm/yr are restricted mainly on the hanging wall block of Rio–Patras fault and specifically towards its NE–termination. Lower values between −1.0 mm/yr and −2.8 mm/yr are observed along the downthrown block of Ag. Triada fault, indicating a lower deformation rate compared to the Rio graben. The footwall block of this fault displays low uplift rates between +0.4 and +1.0 mm/yr. Based on the relative uplift and subsidence velocities on footwall and hanginwall blocks of Ag. Triada fault, a vertical velocity between 1.4 mm/yr and 3.8 mm/yr can be estimated for an almost 10 year period of time. Along the main fault trace and in a 50 m width zone, relatively small ground movements, such as cracks on road pavement and buildings, small scale subsidence and fissures are periodically observed. Typical historic deformation diagrams are shown in Figure 8 for two targets from the footwall and hanging wall of Agia Triada fault.

Pyrgos

Based on the spatial distribution of the PS points, the size and the kind of the vertical movement (uplift or subsidence) for the time period 1992–2000, three areas can be distinguished for the Pyrgos city area: the northern, the central and the southern (Figure 9).

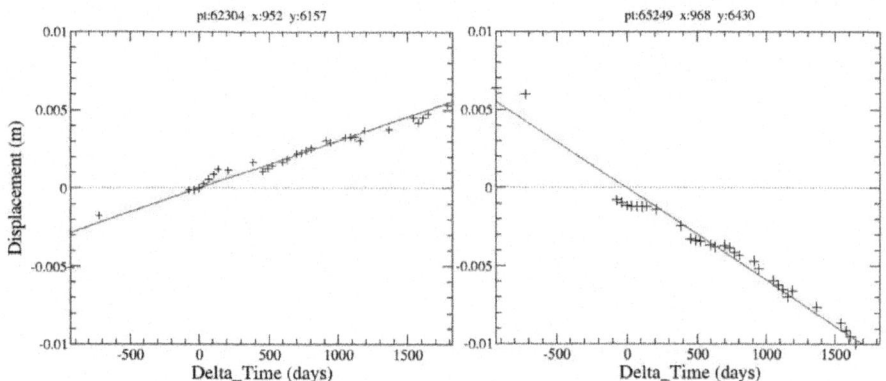

Figure 8: Typical displacement histories obtained by IPTA for specific point targets from the footwall (left diagram) and from the hanginwall (right diagram) of the Agia Triada fault zone.

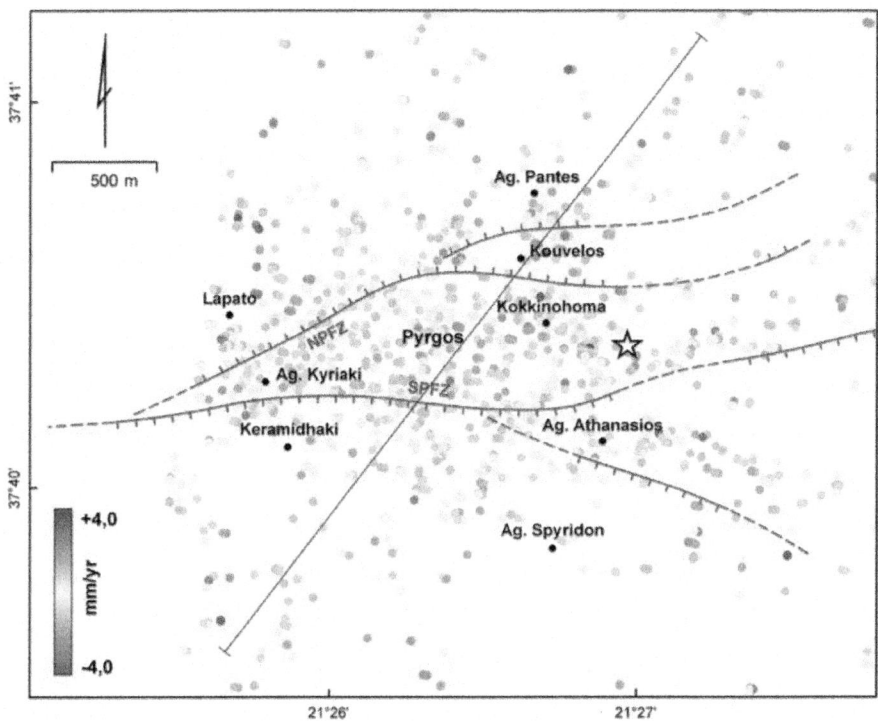

Figure 9: Point targets plotted over the fault map of Pyrgos area, solid line corresponds to the section of Figure 10.

For the northern area the PS points indicate mainly subsidence with values varying from −0.8 to −2.5 mm/yr, while there are few targets, indicating stability. In this area loose alluvial deposits occur, which locally are marshy deposits and the water table of the aquifer is very close to the surface. The area is bounded to the north by the Vounargo fault zone and to the south by the North Pyrgos fault.

The central area, which corresponds to the Pyrgos tectonic horst, is bounded to the north and south by the North and South Pyrgos fault correspondingly [29]. The lower members of the Vounargo formation (thickness >400m, Plio–pleistocene age) occur and the majority of the PS points indicate uplift from +1.5 mm/yr up to +2.5 mm/yr, with very few points presenting subsidence (Figure 10).

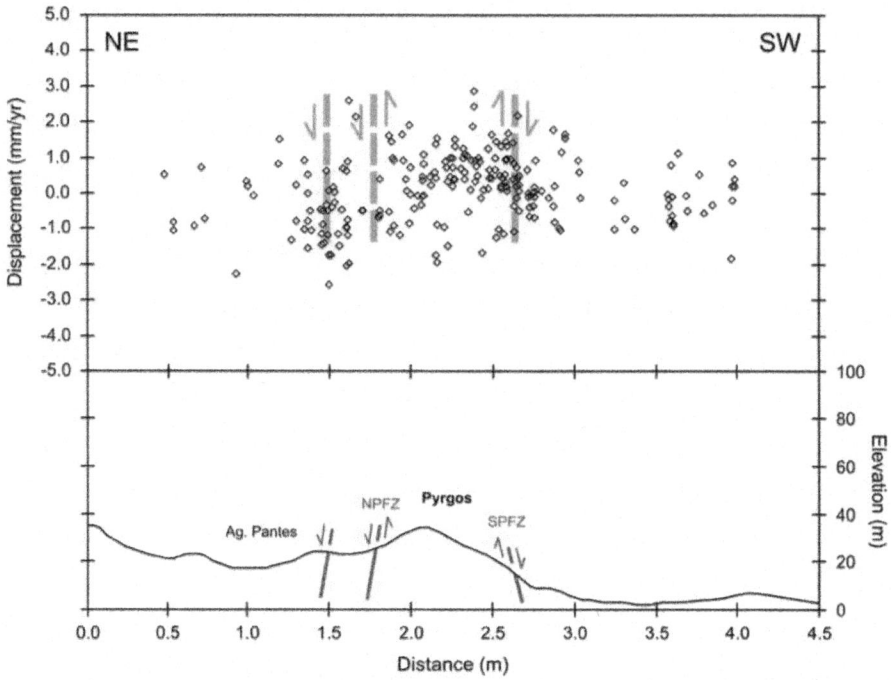

Figure 10: Spatial profile showing the displacement field as observed by IPTA and plotting targets within 150 meters in both sides of the profile (upper), and the related topography (down), red lines correspond to the faults.

At the southern part of Pyrgos, where loose alluvial deposits outcrop, the majority of the targets indicate subsidence, with few being stable and a minority showing uplift.

Typical historic deformation diagram is shown in Figure 11 for a target from the northern part of the city. It has to be mentioned that the measured vertical movements and more specifically the uplift ones are not so intensive. This is in accordance with the low elevation of the Late Pleistocene marine terraces of the broader area, which do not exceed 100 m.

On the contrary, the most destructive earthquakes are related with fault plane solutions of strike slip movements [18,49]. This can explain why the areas of subsidence and uplift are not so clear in the case of Pyrgos being a minor order young neotectonic structure, which has been created after the Middle Pleistocene [29].

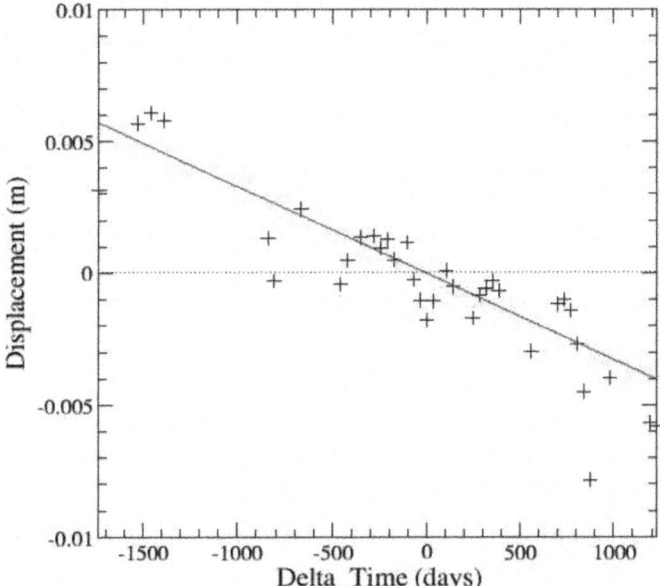

Figure 11: Displacement history obtained by IPTA for specific point target from the northern part of the city.

DISCUSSION

Measurement of ground deformation is a direct and valuable input to models of earthquake risk and for prone areas is of great importance. Given sufficient SAR scenes the submillimeter accuracy of PSI does represent an effective tool for the measurement of tectonic related ground motions.

The PSI technique has proven to be an excellent tool for identifying zones of ground deformation inside urban areas that cannot be easily visible or determined by conventional survey means with unprecedented detail and accuracy. Specifically, the PSI method provides much more information (on both a local and regional scale) on the ground displacements than data obtained by leveling and D–GPS techniques.

Superimposition of several natural and anthropogenic mechanisms that act in different time and scale and provoking ground deformation should be considered. Finally, a complete multidisciplinary approach (including active tectonic, GPS monitoring and seismology) is the appropriate study scheme.

For this specific study, our results of Patras show clearer uplift–subsidence rates due to the more distinct fault pattern and intense deformation compared to Pyrgos area, where more diffused deformation is observed with no significant displacements on surface. This is maybe due to hidden blind structures activating as strike slip at depth, as it is confirmed by recent earthquakes with such focal mechanisms [49,50].

ACKNOWLEDGEMENTS

Data provided by ESA in the frame of CAT–1 projects. The part of the study concerning Pyrgos city was supported by John S. Latsis Public Benefit Foundation and the Patras city case was supported by ESA in the frame of the 1st Call for Ideas for Greek institutions.

REFERENCES

1. Wright, T.; Parsons, B.; Fielding, E. Measurement of interseismic strain accumulation across the North Anatolian Fault by satellite radar interferometry. *Geophys. Res. Lett.* 2001, *28*, 2117–2120.

2. Tosi, L.; Teatini, P.; Carbognin, L.; Brancolini, G. Using high resolution data to reveal depth–dependent mechanisms that drive land subsidence: The Venice coast, Italy. *Tectonophysics*2009.

3. Massonnet, D.; Rossi, M.; Carmona, C.; Adragna, F.; Peltzer, G.; Feigl, K.; Rabaute, T. The displacement field of the Landers earthquake mapped by radar interferometry. *Nature* 1993,*364*, 138–142.

4. Zebker, H.A.; Rosen, P.A.; Goldstein, R.M.; Gabriel, A.; Werner, C.L. On the derivation of coseismic displacement fields using differential radar interferometry. The Landers earthquake. *J. Geophys. Res.* 1994, *99*, 19617–19634.

5. Ferretti, A.; Prati, C.; Rocca, F. Permanent scatterers in SAR interferometry. *IEEE T. Geosci. Remote* 2001, *39*, 8–20.

6. Werner, C.; Wegmüller, U.; Strozzi, T.; Wiesmann, A. Interferometric point target anaysis for deformation mapping. In Proceedings of the IEEE International Geoscience and Remote Sensing Symposium, Toulouse, France, July 2003; pp. 4362–4364.

7. Rosen, P.; Werner, C.; Fielding, E.; Hensley, S.; Buckley, S.; Vincent, P. Aseismic creep along the San Andreas fault northwest of Parkfield, CA measured by radar interferometry. *Geophys. Res. Lett.* 1998, *25*, 825–828.

8. Colesanti, C.; Ferretti, A.; Ferrucci, F.; Prati, C.; Rocca, F. Monitoring known seismic faults using the Permanent Scatterers (PS) technique. In Proceedings of the IEEE International Geoscience and Remote Sensing Symposium, Honolulu, HI, USA, July 2000; pp. 2221–2223.

9.   Lee, J.C.; Angelier, J.; Chu, H.T.; Hu, J.C.; Jeng, F.S.; Rau, R.J. Active fault creep variations at Chihshang, Taiwan, revealed by creep meter monitoring, 1998 – 2001. *J. Geophys. Res-Sol. Ea.*2003, *108*, ETG 4–1– ETG 4–21.

10. Cunha, T.A.; Sarti, F. SAR interferometry as a tool for the detection of active tectonic regions: Preliminary results on the algarve region of the south Portugal. In Proceedings of the FRINGE 2003 Workshop, Frascati, Italy, 1–5 December 2003. ESA SP-550.

11. Taylor, M.; Peltzer, G. Current slip rates on conjugate strike–slip faults in central Tibet using synthetic aperture radar interferometry. *J. Geophys. Res-Sol. Ea.* 2006, *111*, B1240.

12. Biggs, J.; Wright, T.; Lu, Z.; Parsons, B. Multi-interferogram method for measuring interseismic deformation: Denali Fault, Alaska. *Geophys. J. Int.* 2007, *170*, 1165–1179.

13. Cavalié, O.; Lasserre, C.; Doin, M.; Peltzer, G.; Sun, J.; Xu, X.; Shen, Z. Present-day deformation across the Haiyuan fault (Gansu, China), measured by SAR interferometry. In Dragon 1 Programme Final Results 2004–2007, Proceedings of the 2008 Dragon Symposium, Beijing, China, 21–25 April 2008; Lacoste, H., Ouwehand, L., Eds.; ESA Communication Production Office ESTEC: Noordwijk, The Netherlands, 2008; ESA SP-655.

14. Huang, M.; Hu, J.; Ching, K.; Rau, R.; Hsieh, C.; Pathier, E.; Fruneau, B.; Deffontaines, B. Active deformation of Tainan tableland of southwestern Taiwan based on geodetic measurements and SAR interferometry. *Tectonophysics* 2009, *466*, 322–334.

15. Underhill, J.R. Late Cenozoic deformation of the Hellenide foreland, western Greece. *Geo. Soc. Am. Bull.* 1989, *101*, 613–634.

16. Hatzfeld, D.; Pedotti, G.; Hatzidimitriou, P.; Makropoulos, K. The strain pattern in the western Hellenic arc deduced from a microearthquake survey. *Geophys. J. Int.* 1990, *101*, 181–202.

17. Hatzfeld, D.; Kassaras, I.; Panagiotopoulos, D.; Amorese, D.; Makropoulos, K.; Karakaisis, G.; Coutant, O. Microseismicity and strain pattern in northwestern Greece. *Tectonics* 1995, *14*, 773–785.

18. Sachpazi, M.; Hirn, A.; Clément, C.; Haslinger, F.; Laigle, M.; Kissling, E.; Charvis, P.; Hello, Y.; Lépine, J.C.; Sapin, M.; Ansorge, J. Western Hellenic subduction and Cephalonia Transform: local earthquakes and plate transport and strain. *Tectonophysics* 2000, *319*, 301–319.

19. Doutsos, T.; Koukouvelas, I.K.; Xypolias, P. A new orogenic model for the External Hellenides. In *Tectonic Development of the Eastern*

Mediterranean Region; Robertson, A.H.F., Mountrakis, D., Eds.; Geological Society: London, UK, 2006; pp. 507–520.

20. Kokkalas, S.; Xypolias, P.; Koukouvelas, I.K.; Doutsos, T. Post-Ccollisional Contractional and Extensional Deformation in the Aegean Region. Post-Collisional Tectonics and Magmatism in the Mediterranean region and Asia (GSA Special Paper), Dilek, Y., Pavlides, S., Eds.; *Geol. Soc. Am. S.*. 2006, 409, 97–123.

21. Lagios, E.; Sakkas, V.; Papadimitriou, P.; Parcharidis, I.; Damiata, B.N.; Chousianitis, K.; Vassilopoulou, S. Crustal deformation in the Central Ionian Islands (Greece): Results from DGPS and DInSAR analyses (1995–2006). *Tectonophysics* 2007, *444*, 119–145.

22. Hollenstein, Ch.; Müller, M.D.; Geiger, A.; Kahle, H.G. Crustal motion and deformation in Greece from a decade of GPS measurements, 1993–2003. *Tectonophysics* 2008, *449*, 17–40.

23. Doutsos, T.; Pe-Piper, G.; Boronkay, K.; Koukouvelas, I. Kinematics of the Central Hellenides.*Tectonics* 1993, *12*, 936–953.

24. Xypolias, P.; Doutsos, T. Kinematics of rock flow in a crustal scale shear zone: implication for the orogenic evolution of the SW Hellenides. *Geol. Mag.* 2000, *137*, 81–96.

25. Skourlis, K.; Doutsos, T. The Pindos Fold and Thrust Belt (Greece): Inversion kinematics of a passive continental margin. *Int. J. Earth Sci.* 2003, *92*, 891–903.

26. Sotiropoulos, S.; Kamberis, E.; Triantaphyllou, M.; Doutsos, T. Thrust sequences at the central part of the External Hellenides. *Geol. Mag.* 2003, *140*, 661–668.

27. Doutsos, T.; Kontopoulos, N.; Poulimenos, G. The Corinth–Patras rift as the initial stage of continental fragmentation behind an active island arc (Greece). *Basin Res.* 1988, *1*, 177–190.

28. Doutsos, T.; Kontopoulos, N.; Frydas, D. Neotectonic evolution of northwestern continental Greece. *Geol. Rundsch.* 1987, *76*, 433–452.

29. Lekkas, E.; Papanikolaou, D.; Fountoulis, I. The Pyrgos Earthquake: The geological and geotechnical conditions of Pyrgos area (W. Peloponnese, Greece). In Field–guide for the Pre–Congress Excursion of the XV Congress of the Carpatho–Balcan Geological Association, Athens, Greece, September 1995; pp. 42–46.

30. Koukouvelas, I.; Mpresiakas, A.; Sokos, E.; Doutsos, T. The tectonic setting and earthquake ground hazards of the 1993 Pyrgos earthquake, Peloponnese, Greece. *J. Geol. Soc. London*1996, *153*, 39–49.

31. Doutsos, T.; Kokkalas, S. Stress and deformation patterns in the Aegean region. *J. Struct. Geol.* 2001, *23*, 455–472.

32. Papanikolaou, D.; Fountoulis, I.; Metaxas, C. Active faults, deformation rates and Quaternary paleogeography at Kyparissiakos Gulf (SW Greece) deduced from on-shore and off-shore data. *Quatern. Int.* 2007, *171–172*, 14–30.

33. Zelilidis, A.; Koukouvelas, I.; Doutsos, T. Neogene paleostress changes behind the forearc fold belt in the Patraikos Gulf area, Western Greece. *Neues Jahrb. Geol. P. M.* 1988, 311–325.

34. Flotte, N.; Sorel, D.; Muller, C.; Tensi, J. Along strike changes in the structural evolution over a brittle detachment fault: example of the Pleistocene Corinth – Patras rift (Greece). *Tectonophysics* 2005, *403*, 77–94.

35. Chronis, G.; Piper, D.W.; Anagnostou, C. Late Quaternary evolution of the Gulf of Patras, Greece: Tectonism, deltaic sedimentation and sea–level change. *Mar. Geol.* 1991, *97*, 191–209.

36. Koukis, G.; Sabatakakis, N.; Tsiambaos, G.; Katrivesis, N. Engineering geological approach to the evaluation of seismic risk in metropolitan regions: Case study of Patras, Greece. *Bull. Eng. Geol. Environ.* 2005, *64*, 219–235.

37. Kalteziotis, N.; Koukis, G.; Tsiambaos, G.; Sabatakakis, N.; Zervogiannis, H. Structural damage in a populated area due to an active fault. In Proceedings of the 2nd International Conference on Recent Advancces in geotechnical Earthquake Engineering and soil Dynamics, Rolla, MO, USA, 11–15 March 1991; pp. 1709–1716.

38. Hasiotis, T.; Papatheodorou, G.; Kastanos, N.; Ferentinos, G. A pockmark field in the Patras Gulf (Greece) and its activation during the 14/7/93 seismic event. *Mar. Geol.* 1996, *130*, 333–344.

39. Papazachos, B.C.; Papazachou, C. *The Earthquakes of Greece*; Ziti Publications: Thessaloniki, Greece, 1997.

40. Papatheodorou, G.; Christodoulou, D.; Geraga, M.; Etiope, G.; Ferentinos, G. The pockmark field of the Gulf of Patras: An ideal natural laboratory for studying seabed fluid flow. In Filed trips guide book "Sedimentology of western and central Greece from Triassic to recent", Proceedings of the 25th IAS Meeting of Sedimentology, Patras, Greece, 4–7 September 2007.

41. Lepichon, X.; Angelier, J. The Hellenic arc and trench system: a key to the neotectonic evolution of the Eastern Mediterranean area. *Tectonophysics* 1979, *60*, 1–42.

42. Mariolakos, I.; Papanikolaou, D. The neogene basins of the Aegean Arc from the Paleogeographic and the Geodynamic point of view. In Proceedings of the Int. Symp. Hell. Arc and Trench (HEAT), Athens, Greece, 1981; Vol. I, pp. 383–399.

43. Mariolakos, I.; Papanikolaou, D. Deformation pattern and relation between deformation and seismicity in the Hellenic arc. *Bull. Geol. Soc. Greece* 1987, *XIX*, 59–76. (in Greek).

44. Mariolakos, I.; Papanikolaou, D.; Lagios, E. A neotectonic geodynamic model of Peloponnesus based on: morphotectonics, repeated gravity measurements and seismicity. *Geol. Jb.* 1985, *50*, 3–17.

45. Kamberis, E. Geology and petroleum geology study of NW Peloponnese, Greece. Ph.D. Thesis, National Technical University of Athens, Athens, Greece, 1987. (in Greek).

46. Lekkas, E.; Papanikolaou, D.; Fountoulis, I. Neotectonic map of Greece, sheet "Pyrgos" – "Tropaia" (scale 1/100.000). Project-University of Athens, Athens, Greece, 1992, 120. (in Greek).

47. Lekkas, E.; Fountoulis, I.; Papanikolaou, D. Intensity distribution and Neotectonic macrostructure Pyrgos Earthquake data (Greece, 26 March 1993). *Nat. Hazards* 2000, *21*, 19–33.

48. Scharoo, R.; Visser, P.N.A.M. Precise orbit determination and gravity field improvement for the ERS satellites. *J. Geophys. Res.* 1998, *103*, 8113–8127.

49. Roumelioti, Z.; Benetatos, Ch.; Kiratzi, A.; Stavrakakis, G.; Melis, N. A study of the 2 December 2002 (M5.5) Vartholomio (western Peloponnese, Greece) earthquake and of its largest aftershocks. *Tectonophysics* 2004, *387*, 65–79.

50. Koukouvelas, I.; Kokkalas, S.; Xypolias, P. Surface deformation during the Mw 6.4 (8 June 2008) Movri earthquake in the Peloponnese and its implications for the seismotectonics of Western Greece. *Int. Geol. Rev.* 2009.

Chapter 7

LAND USE CHANGES AND ENVIRONMENTAL PROBLEMS CAUSED BY BANK EROSION: A CASE STUDY OF THE KOLUBARA RIVER BASIN IN SERBIA

Slavoljub Dragicevic[1], Nenad Zivkovic[1], Mirjana Roksandic[1], Ivan Novkovic[1], Stanimir Kostadinov[2], Radislav Tosic[3], Milomir Stepic[4], Marija Dragicevic[5] and Borislava Blagojevic[6]

[1]University of Belgrade, Faculty of Geography, Belgrade, Serbia

[2]University of Belgrade, Faculty of Forestry, Belgrade, Serbia

[3]Faculty of Natural Sciences, Banja Luka, Republic of Srpska

[4]Institute for Political Studies,Belgrade, Serbia

[5]First Elementary School in Obrenovac, Obrenovac, Serbia

[6]University of Nis, Faculty of Civil Engineering and Architecture, Serbia

INTRODUCTION

Geomorphological analysis of the dominant erosion processes and their intensity quantification were done in the previous researches of the Kolubara River basin [1-3]. The results showed that, the level of the landscape degradation and modification of geomorphologic processes by human activities has been increased in the past decades [4], and it was initiated by very fast demographic, socio-economic and technological changes in Serbia, likewise in the region [5-7], and in the world [8-11].

According to level and type of degradation, the Kolubara River basin belongs to the most endangered areas in Serbia. Due to the lignite exploitation in the Kolubara River basin, human impact led to morphological change of the entire area, as well as to the changes of the intensity of different geomorphologic processes: changes in river course [12,13], the intensity of bank erosion [14,15], sediment deposition [16] and environmental problems [17,18].

Unlike the other rivers with similar hydrological characteristics, the river network in the lower part of the Kolubara River basin were changed rapidly during the XX century because of direct human impact. Anthropogenic influences on the hydrological network in the study area were very intensive since 1959, when the huge river regulation works were done in the lower part of the Kolubara River. Spatial planning of the area, which included diverting of the Kolubara's river bed, had an aim to prepare the site for the lignite exploitation within the Kolubara mining basin. The Kolubara River divides the mining basin in two parts: eastern and western part. The productive area of the basin (geologic contours of lignite deposits) is 520 km². Kolubara mining basin is situated about 40 km south-southeast of Belgrade and represents the largest lignite deposits in the central part of Serbia; the annual production is 30 million tons of lignite, and it is the opencast mine. The mine expansion caused the need for technical solutions of diverting and removing river beds in this area. According to "General project of diverting the Kolubara River and its tributaries for the purpose of lignite exploitation", the Kolubara's riverbed was diverted into the Pestan's riverbed (its right tributary). This caused many problems which were not predicted by the General project.

In this way, anthropogenic factor modified existing natural conditions: the process of fluvial erosion was changed; bank erosion became stronger and resulted in soil loss, larger amounts of sediment load deposition, cutting off the meanders and fossilization of certain parts of the riverbed, floods, land use changes, landscape degradation, sediment load pollution, etc.

RESEARCH AREA

Regarding to natural conditions, the Kolubara River basin is similar to the other river basins in the area. Tectonic movements had an influence on a morphological evolution of the river network in the past. During the Paleogene and the Early Neogene a small bay of the Pannonian Sea named the Kolubara's bay existed in the area of the Kolubara River basin. After the sea recession, the fluvial erosion started in this bay and it formed today's hydrological network of the Kolubara River. Tectonic characteristics of this area, more precisely Kolubarsko-pestanski fault and Posavski fault had influenced the orientation of the hydrological network in the Kolubara River basin. But today the Kolubara's hydrological network is influenced by fluvial erosion and anthropogenic factors.

The Kolubara River Basin encompasses the western part of Serbia and covers 4.12% of Serbia's surface area. The highest point of the drainage basin is at 1,346 m, and the lowest has altitude of 73 m. The Kolubara River is the last large right tributary of the Sava River, and according to the flow length (86.4 km) and the basin area (3,641 km²) it is classified as a middle-sized river on the territory of Serbia [3]. The lower part of the Kolubara River basin is called the Donjokolubarski basin (area of 1,810 km²) and is situated in the municipality of Obrenovac. The Donjokolubarski basin encompasses the catchment area of the Kolubara's confluences (the Pestan River, the Turija River with the Beljanica River, the Tamnava River with the Ub River and the Kladnica River) and the lower part of the Kolubara's valley. The average altitude of the Donjokolubarski basin is 168 m, the highest point is at 695 m, and the lowest has an altitude of 73 m.

According to the nearest meteorological station in Obrenovac, this area is characterized by continental climate, the average temperature was 11°C, and the mean annual precipitation from year 1925 to 2000 was 722 mm [12]. The average annual runoff of the Kolubara River (at Drazevac gauging station) for the period 1961-2005 was 21.8 m³/s.

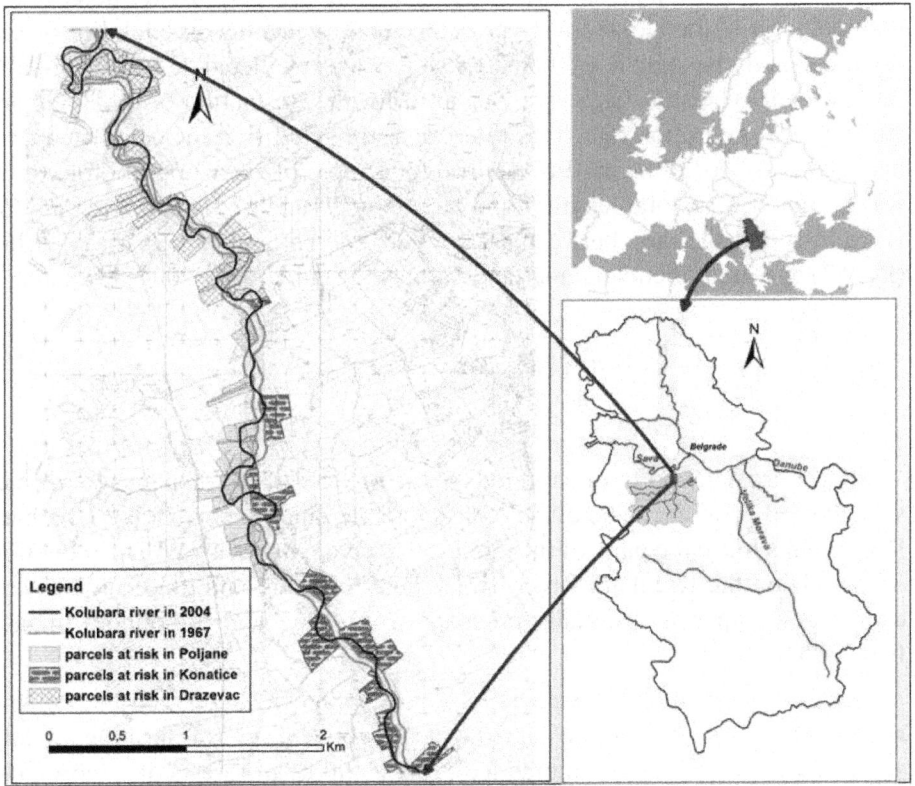

Figure 1: Position of the Kolubara River basin in Serbia (right) and study area (left)

METHODOLOGY

In this research we used diffrent methods that can be devided into the field and lab work methods. The GIS methods were used for the modeling of terrain evolution and landscape changes, which represents the base for bank erosion intensity quantification.

Analysis of topographical maps, aerial photo and orthophoto images were used in the previous researches aiming to determine the evolution of the riverbed [6,15,19-23]. The results showed that the application of GIS has an advantage in quantification of river migration processes.

For the purposes of this study, comparative analyses have been made on the base of Cadastral maps scale 1:2500 from 1967 and orthophoto images from 2004; reconstruction of the hydrological system has been done for the periods from 1967 to 2004. By comparing the data from two periods, we determined the evolution of the Kolubara River course in 37 years. River bank lines were digitized and the extent of bank erosion was calculated under Geomedia professional. The same software was used for the estimation of the Kolubara River lateral migration rate. This rate was estimated using the calculated area between river positions in 1967 and in 2004 (area of river migration), which was divided by the total length of the river course in 1967. The loss of land (S) is expressed as the ratio between area of endangered land parcels (ha) in 1967 (P1967) and area of endangered land parcels (ha) in 2004 (P2004) [15]:

$$S = \frac{P_{1967} - P_{2004}}{P_{1967}} * 100$$

River erosion and frequent floods make great material damages to people, villages and economy. The owners of the arable land parcels on the Kolubara River banks loose the parts of the parcels that river carries away. The reduction of parcels on the Kolubara River banks, land loss and land use changes were estimated comparing the cadastral maps from 1967 and orthophoto images from 2004.

Land use structure in the area of villages: Drazevac, Konatice and Poljane are characterized by: arable land (which people used for farming mostly wheat and corn-crop rotation practice), forests (alluvial forests of willows and poplars) and few pastures. The river dynamic is intensive in the Kolubara's alluvial zone, which influenced sandbank formation, mostly on the concave side of the river. By statistical analysis of a land use structure [24] in the three villages with degraded land parcels on the river banks, we obtained the results which show significant reduction in arable land. And by analyses of the questionnaire carried out among the owners of degraded land parcels in the villages Drazevac, Konatice and Poljane, it can be concluded that it was significant decrease in the agricultural production. The risks from the floods and further soil loss influenced the land owners' decision making about farming the degraded land parcels.

The change of fluvial erosion intensity was analyzed regarding to changes in water balance and sediment load transport on two hydrological profiles. The results of water balance that D. Dukic [25] has made in his research for the period of 1925-1960 and the results obtained in this study were analyzed and compared. This comparative analysis appoints to the amount of water which Donjokolubarski basin disposed before regulatory changes of Kolubara in 1959/60 and after them. River flow regimes of different periods were compared because that could be a factor which has a significant influence on the observed process. All these efforts should confirm or eliminate the influence of natural factors on the river banks degradation in the Donjokolubarska valley.

Having data of extreme discharges, in order to estimate the impact of future floods on bank erosion, we have made a probability curve of maximum discharges of the Kolubara River and its tributaries.

Because of intense anthropogenic impacts in the Donjokolubarska valley, we have sampled the suspended sediments from the Kolubara's riverbed and later analyzed the pollution of the accumulated load. Since the processes of bank erosion and sediment accumulation occur close to the villages and that endangered land parcels are used for food production, such approach points to ecological aspect of researched problem.

The sediment samples were taken on two locations in the Kolubara's riverbed. For heavy metals and carbon analysis the soil was milled to a fine powder. Heavy metals were determinate by AAS method.

THE INTENSITY OF BANK EROSION

Natural Conditions Changes as a Factor of Bank Erosion in the Study Area

On the research sector (Fig. 1) the Kolubara River length in 1967 was 8.2 km and 10.6 km in 2004. This fact appoints to the river course evolution through the landscape. In the period between 1967 and 1981 the Kolubara River has migrated 50 m, actually 27 m into left and 23 m into right, and the average migration of the Kolubara River was 3.6 m per year. By further comparison of aerial photo image from 1981 and orthophoto image from 2004 it can be observed that the Kolubara's riverbed was stabilized and during 23 years migrated only 26 m. So, the Kolubara River average migration in this period was 1.1 m per year which is three times less comparing to the previous period of observation (1967-1981) [13].

The rate of the Kolubara river lateral migration along the research sector is 47 m in average for the period of 37 years, which means 1.27 m per year. At the most endangered part (in the area of Drazevac village) the most intensive migration rate of the riverbed was 224 m in 37 years, with the average of 6.05 m per a year [15].

The changes of fluvial erosion intensity may result from changes in climatic-hydrological characteristics of the river basin (which are manifested in discharge regime changes) and various human impacts. Therefore, the natural factors of the Donjokolubarski basin were analyzed to determine whether they have influenced the stronger bank erosion.

The results showed that average mean annual discharge of the Kolubara River measured in Drazevac was 22.3 m^3/s in the observation period 1961-1990, and 21.3 m^3/s in the observation period 1991-2005. Amplitude of average high and low flows in the period 1961-1990 was 77.94 m^3/s, and in the period 1991-2005 it was 64.66 m^3/s [13].

To study water balance of the Donjokolubarski basin we used the following periods: 1925-1960, 1961-1990 and 1991-2005. With this approach it was possible to determine the changes that may be occurred after diverting the Kolubara River into the Pestan's riverbed in 1959. Briefly, precipitation analysis showed that the second period was a bit wetter than the first, actually about 60 mm in the Pestan River basin and 80 mm in the Turija River basin and the Tamnava River basin. Meanwhile, higher air temperatures and higher evaporation caused almost the same specific discharges of these rivers. The last period was in mean values similar to the second, apart from intensified variation of extreme values of all climatic elements. The discharges were influenced by more frequent alternation of wet and dry periods, which could be seen on figure 2.

Monthly coefficients of variation of the period 1991-2005 are higher in all river sub basins except in July and August. These differences are significant, the variation of discharges in eight months are higher than the highest coefficients of variation of the period 1961-1990, which is 1.5. The more important is the fact that the period of appearances of unstable discharges is March-April (over 2.5), which is related to snow melting. That is the period of maximum discharges and any sudden disturbance of soil moisture resulting in serious disorder of river bank stability. Some of the natural factors have been changed in the last two decades, for example, March used to be, in Serbia, the month with the most stable discharges (and the highest). The differences in March discharges during the observed periods are over 20 m³/s (almost twice reduced), and it was followed by extreme discharge variations. These are the significant changes since the area of the Kolubara River basin is bigger than 3500 km²and maximum discharges are higher than 500 m³/s. Although the mean values are not of crucial importance, they indicate some disturbances which should be kept in mind; particularly because the last period of observation is twice shorter than the previous one and all analyses in the world indicate that extreme values of natural phenomena are more pronounced and more frequent.

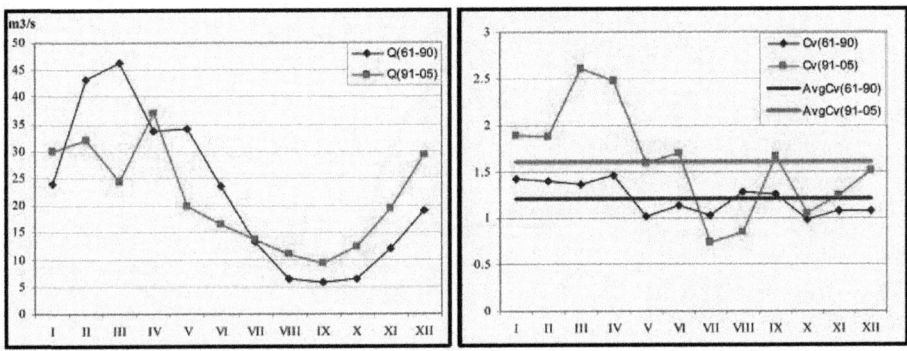

Figure 2: Mean monthly discharges (Q) and coefficients of variation (Cv) of the Kolubara River measured in Drazevac gauging station for the both periods of observation

Preliminary results of the Donjokolubarski basin annual flow variation show discrepancy in spring and summer monthly flow among two studied periods (Figure 2 - right). Further research should examine correlation of monthly flow and bank erosion intensity.

Anthropogenic Influences as a Factor of Bank Erosion in the Study Area

The erosion control works in the river channel can cause changes in river morphology, since they influence the changes in river regimes, river bank characteristics and amount of sediment transport [26]. The consequences of these interventions are numerous, and often lead to riverbed widening and undermining concave sides of the river banks. The processes of river bank collapsing and erosion are complex since they are results of several factors, including sediment transport, ground lithology, stratigraphy, slope, flow geometry and anthropogenic activities [27].

Opencast lignite exploitation in Kolubara mining basin started in 1952 when the mining field "A" was open (it was exploited till 1966). Mining field "B" was opened in 1952, mining field "D" in 1961, "Tamnava-East Mining Field" in 1979, "Tamnava-West Mining Field" in 1994 and mining field "Veliki Crljeni" in 2008 [28].

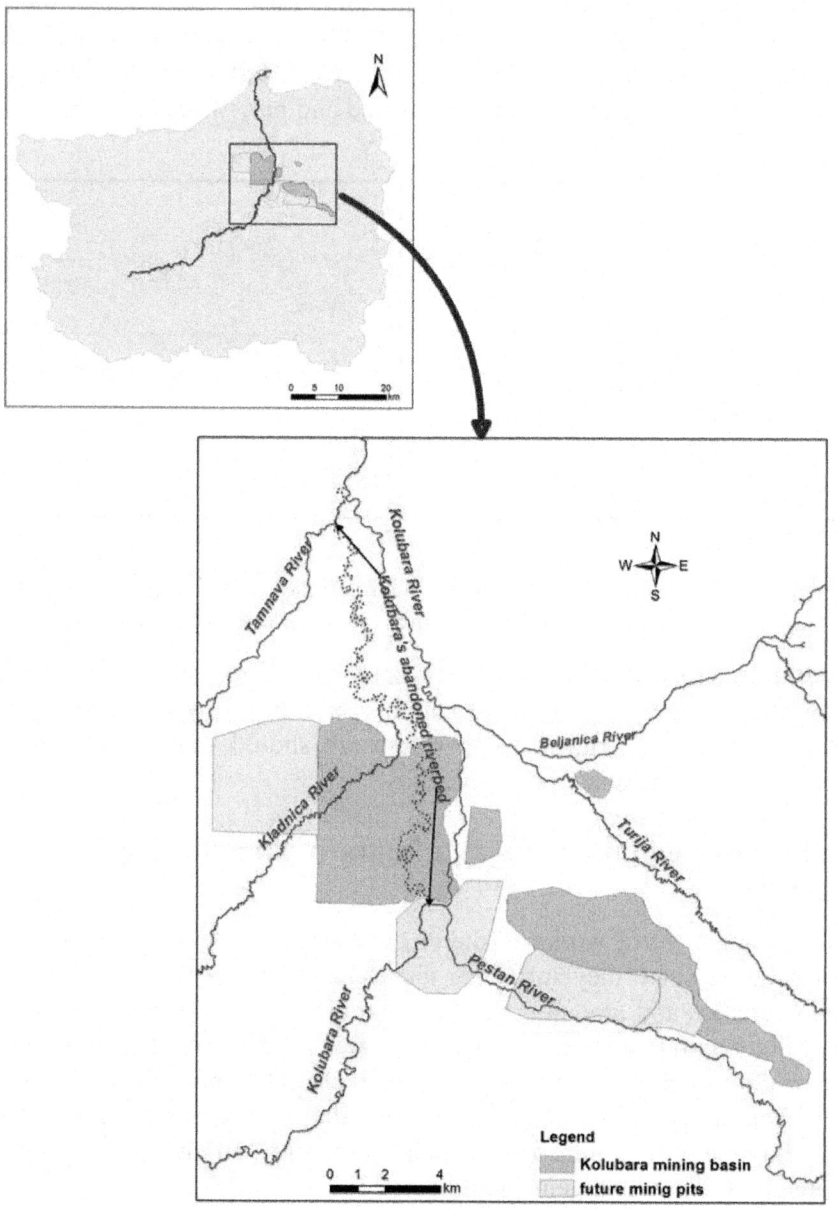

Figure 3: Plan of the mining fields in Kolubara mining basin [28]

Beogradsko-posavska water community "Beograd" has made a project „Regulation of the Kolubara River and its right tributaries from Ćelije to Poljane (km 23+200 – km 55+506)" in 1957. Regulation works on the Kolubara River and its tributaries had begun in 1959/60. The diverting of the Kolubara River was done to clear the area for lignite exploitation, actually for opening new mining fields.

From that moment the Kolubara River flows through the Pestan›s riverbed (its right tributary), and previous Kolubara›s riverbed is abandoned with the periodic flow. The length of the Kolubara River was shortened by 20 km because of diverting its riverbed, while the length of the Pestan River was also shortened because its confluence was moved to the South. By diverting the Kolubara›s riverbed into the Pestan›s riverbed, which morphologically was not predisposed for kinetic energy of stronger flow, bank erosion became a dominating geomorphological process in the area and initiated processes of digging the riverbanks, transportation and deposition of eroded material. It is obvious that river system changes in lower part of the Kolubara River are demonstrated in domination of fluvial (lateral) erosion on one hand, and in cutting the meanders and fossilization of certain parts of the riverbed, on the other hand.

THE CONSEQUENCES OF BANK EROSION

Forming of Meanders

Map of the Kolubara's basin first trend of relief energy [2] shows that almost whole area of the Donjokolubarska valley is under tectonic movements of slowly sinking. For this reason the sediments are accumulated in the riverbed, river velocity decreases which cause the riverbed meandering and stronger bank erosion. This natural process became more intensive since the Kolubara River was diverted into the riverbed of Pestan. In the Donjokolubarski basin there are numerous sectors with abandoned riverbeds and cut off meanders.

Forming of meanders and cutting the "necks" are recent geomorphologic-hydrological process, which is dominated in the study area. According to results of the recent researches [13], there are 89 abandoned parts of the riverbeds and cut off meanders in the area of the Donjokolubarski basin. The Kolubara and its tributaries tend to move to the east because of the Kolubarsko-pestanski fault, which indicate more abandoned riverbeds and cut off meanders on the left side of the Kolubara valley (64), compared to the right side (25).

In the study area there are 40 cut off meanders with total length of 20.30 km while the number of abandoned parts of the riverbeds is 49 with total length of 76.03 km. Hence, the total length of all abandoned riverbeds and cut off meanders in the Donjokolubarski basin is 96.33 km, and their total surface is 3.35 km². The longest cut off meander is 1.7 km long and the shortest is 185.7 m long. The longest abandoned riverbed is 6.49 km long.

The length of the Kolubara's riverbed is influenced by stronger bank erosion and formation of meanders, which is clearly perceived in the field. According to orthophoto image from 2004 and satellite image (Google Earth) the Kolubara River length (in the Donjokolubarska valley) is 66.52 km, while according to topographical map from 1970 it was 67.5 km, and according to topographical map from 1925 it was 87.6 km.

After cut off meander, the riverbed itself morphologically adjusts to the new state [29]. Morphological changes of the rivers are reflected in digging the concave river banks and sediment accumulation on the convex river banks.

Land Use Changes

As we earlier indicated, river erosion and frequent floods can make great material damages to people, villages and economy. Since the lateral erosion has more intensity, the river banks on concave side of the Kolubara River often collapse and farmers who have arable land parcels on the river bank (in the area of three villages the Kolubara flows through) loose the parts of the parcels which were carried away by the river. Based on Cadastral maps from 1967 and orthophoto images from 2004 we have estimated the area of diminished land parcels and their land loss.

Farmers who have land parcels in three villages (Drazevac, Konatice and Poljane) on the Kolubara river bank cannot farm them in whole, because the river has changed its course and took some parts of the land parcels away. The cadastral maps of the researched area scale 1:2500 from 1967 and orthophoto images from 2004 were compared. Using the results of this comparative analysis, the evolution of the hydrological system in the period from 1967 to 2004 was presented. The previous research showed that 60.37 ha was lost and degraded by the river bank erosion, which means that the land loss is 50.57 % of the land parcels from 1967 [15].

Table 1: Land use structure in Drazevac, Konatice and Poljane

Land use	Total	Arable land	Woods	Pastures	Mead-ows	Sand banks	Other (roads…)
Drazevac							
number of end-agered parcels	95	39	16	1	5	34	-
area in 1967 (ha)	57.76	34.96	5.56	0.25	2.93	14.07	-
area in 2004 (ha)	28.23	21.76	2.48	0.02	1.30	2.67	-
loss of land (ha)	29.53	13.20	3.08	0.23	1.63	11.40	-
Konatice							
number of end-agered parcels	86	56	2	2	-	25	1
area in 1967 (ha)	50.44	42.73	0.49	1.02	-	6.12	0.09
area in 2004 (ha)	32.13	29.57	0.34	0.55	-	1.60	0.06
loss of land (ha)	18.31	13.16	0.15	0.47	-	4.52	0.03
Poljane							
number of end-agered parcels	66	41	3	-	-	21	1
area in 1967 (ha)	40.09	33.91	0.82	-	-	5.11	0.25
area in 2004 (ha)	26.26	25.15	0.19	-	-	0.89	0.03
loss of land (ha)	13.83	8.76	0.63	-	-	4.22	0.22

On the basis of the recent and more accurate data from Obrenovac Municipality Cadastre we have determined land use structure of degraded land parcels on the Kolubara River banks. According to these data, total area of all 247 endangered land parcels was 148.3 ha in 1967, and 86.62 ha in 2004. Therefore, 61.68 ha of soil were lost within 37 years [13].

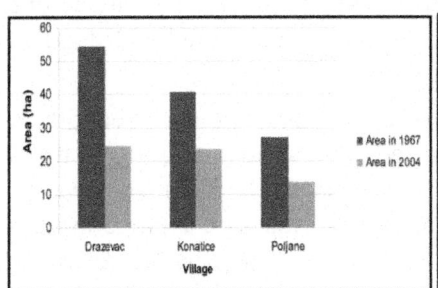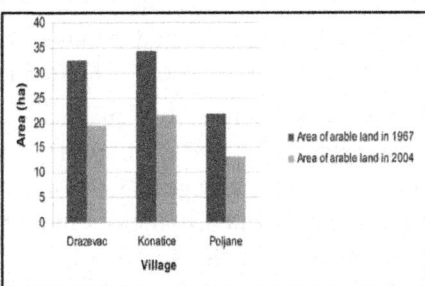

Figure 4: Land use changes in total area (left) and area of arable land (right) in Drazevac, Konatice and Poljane between 1967 and 2004

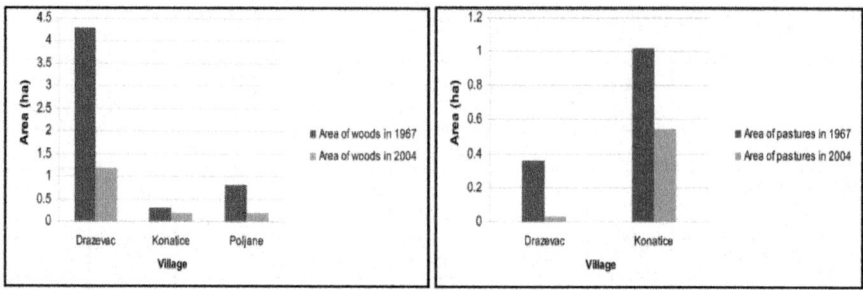

Figure 5: Land use changes in area of woods (left) and area of pastures (right) in Drazevac, Konatice and Poljane between 1967 and 2004

From 247 endangered land parcels, 136 are arable land with the area of 111.6 ha in 1967, and 76.48 ha in 2004, which means that within 37 years 35.12 ha of arable land was lost for farming, and it is 31.5 % of the initial area (in 1967). The woods comprise 21 of all endangered land parcels with area of 6.87 ha in 1967, and 3.01 ha in 2004, which means that it has been lost 3.86 ha of woods. There are only the three endangered land parcels with pastures, and their area was 1.27 ha in 1967, and 0.57 ha in 2004. All five endangered land parcels with meadows are in the area of Drazevac village.

Analyzing the area of endangered parts in the three villages, one can conclude that erosion was the most intensive in the period 1967-1981, when 50.9 ha of soil was lost within 14 years. The riverbed was stabilized later and the erosion decreased. This appoints to the fact that diverting the Kolubara River into the Pestan's riverbed caused more intensive bank erosion since in time erosion was diminished which brought to the riverbed stabilization.

Three villages on the Kolubara River banks (Drazevac, Konatice and Poljane) were characterized by agricultural production and agricultural population. Analyzing the land use structure of endangered parcels one can conclude that arable land parcels are the most endangered and degraded by intensified lateral erosion of Kolubara.

In Serbia there is 4.25 million ha of arable land, and each year 500000 ha (which means 11.74 %) of arable land remain uncultivated [30]. In the above mentioned three villages 33.47 % of arable land (on the Kolubara River banks) remain uncultivated, which is three times more than the average in Republic of Serbia. During the field work, the interviewed owners of endangered arable land parcels pointed that they do not farm their land on the river banks because of flood risks. The Kolubara River floods almost every year and crop is ruined. Therefore, besides the loss of arable land, frequent floods are huge problem in this area.

The economic consequences of bank erosion in the area of the Donjokolubarski basin could be analyzed through losses that the owners of endangered arable land parcels had (because the arable land parcels were reduced). The area of arable land (on the river banks) was diminished by 35.12 ha within 37 years. In the research area the average annual yield is 3-4 t per hectare, so the annual losses of crops (mostly wheat and corn) in recent years are between 100 and 140 t per year.

Sediment Load Discharge

Changes of land use structure and changes in sediment regimes are the direct consequences of the bank erosion [22]. The calculation of one-day sediment load discharge at the monitored hydrological profile includes the values of mean daily flow ($Q - m^3/s$) and the relevant concentration of the suspended load ($C-mg/l$). The assessment of sediment deposition rate is based on the results of RHMSS [31] measurements and the results of own daily measurements of suspended load concentration during the period (1985-2004). The results show that 193253.8 tons of material was accumulated between two hydrological profiles. And the riverbed itself was raised for 36 cm, which is nine times enlarged comparing to previous research when it was raised for 4.2 cm (with a shorter time series) [2]. The extraction of the river deposited sediments from the Kolubara's riverbed was stopped. Although the river deposits were hand extracted with low intensity, it certainly had great positive effects from the aspect of maintaining the surface of riverbed profile. Simple solutions, like the river deposit extraction, do not need huge investments for the implementation and they can be carried out without limitation of the natural conditions.

REIK Kolubara has a negative ecological impact on the Donjokolubarska valley. There are lots of waste waters after the ore production. Waste waters from the mine "REIK Kolubara" are discharged without any treatment into the Kolubara River. Therefore, the Kolubara River contains waste waters from the mine and after each flood the soil on the river banks is contaminated by substances from the waste waters. The results of soil analysis in the Kolubara river basin show increased concentration of nickel, arsenic and lead in the area of the Donjokolubarski basin [18].

The eroded material from the river banks is accumulated downstream. The accumulated sediments can contain considerable concentrations of heavy metals and that is threat for the aquatic habitats and for the people [8, 32-34].

Ecological aspects of mechanical water pollution by suspended sediment, chemical water pollution by organic and mineral fertilizers used in plant production in the catchment, nutrients found in the soil as well as chemical pollution of water and sediment by pesticides and heavy metals are very important ecological problem in the study area. On two locations we have sampled the accumulated material from the Kolubara's riverbed to examine the transport of contamination and accumulation of contaminated sediments due to bank erosion processes.

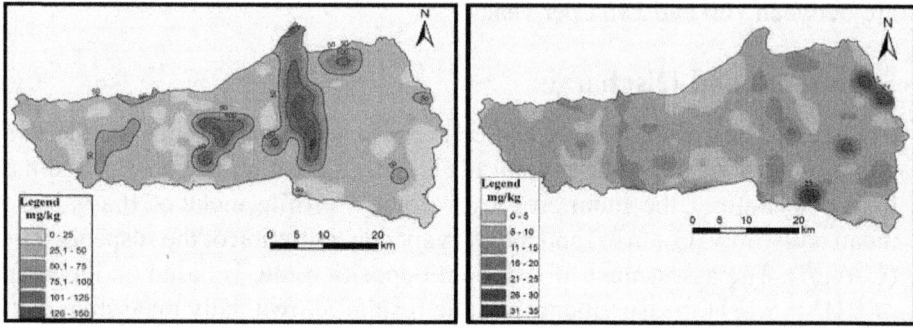

Figure 6: Distribution of Ni (left) and As (right) in the soil of the Donjokolubarski basin [18].

Figure 7: Sampling of deposited sediments on location 1 (left) and 2 (right)

The deposited sediments have sandy-clay texture. Chemical characteristics of deposited sediments from Kolubara's riverbed are: mildly alkaline reaction, high bases saturation degree and low humus content. The average heavy metal concentration in sediments decreased in the order: Ni > Cr > Zn > Pb > Cu > Cd.

In Serbia there is no law defining limitation of heavy metals in suspended sediments. Some European countries have such laws [34], but the differences between countries are significant. In most of the cases the critical values are obtained using the equilibrium method and maximum acceptable concentration

(MAC) for the surface waters with regard to direct and indirect effects on living organisms in the water-sediments systems. According to these data, the range for different elements is as follows: 15 - 100 mg.kg^{-1} for Pb; 0.6 – 2.4 mg.kg^{-1} for Cd; 36 - 120 mg.kg^{-1} for Cu; 123 - 1050 mg.kg^{-1} for Zn; 10 - 180 mg.kg^{-1} for Ni and 37 - 120 mg.kg^{-1} for Cr. These ranges are bigger then estimated ecotoxic criteria [35], which are: 5 - 50 mg.kg^{-1} for Pb, Ni i Cu; 0.1 – 1.0 mg.kg^{-1} for Cd; 50 - 500 mg.kg^{-1} for Zn; 10 - 100 mg.kg^{-1} for Cr.

Table 2: Heavy metal contents in the deposited sediment load

Sample	Zn	Cu	Pb	Cd	Cr	Ni
	mg.k^{g}1					
Location 1	39.3	15.9	23.0	0.0	94.0	198.3
Location 2	41.0	20.2	27.9	0.1	103.0	210.9

Respecting the above mentioned criteria, mean measured concentration of Pb, Zn and Cr are within the limits (after de Vries and Bakker [35]), while concentration of Cu and Cd are below the limits and concentration of Ni are above the limits. According to OSPAR limitation values [36], average concentrations of Pb, Cu and Cd are below the limits, average concentrations of Cr and Zn are within the limits, while average concentrations of Ni are above the limits.

Floods

The Kolubara River is a good example which represents the existence of all conditions for frequent and large scale floods. As an indirect consequence of the anthropogenic influence on the hydrological system in the lower part of the Kolubara valley, once a year (sometimes twice a year) the Kolubara River overflows, and the area of lower part of the Kolubara River basin is endangered by floods. Catastrophic floods of the Kolubara River and its tributaries spread over the area of lower part of the Kolubara River basin during the spring of 1937, and they lasted two months approximately (from March to May). In this area large scale floods also happened in 1965, 1975, 1981, 1996, 1998, 1999, 2001, 2004, 2006, 2008 and 2010.

The highest discharge of the Kolubara River in the period of 1959-2000 was 646 m^{3}/s and it was registered on Drazevac hydrological station. According to probability curve of high discharges the discharge of 646 m^{3}/s may occur once in a 46 years. The lowest value of annual maximum discharge would be about 25 m^{3}/s, the highest discharge in a hundred years would be 740 m^{3}/s, and the highest discharge in a thousand years would be 960 m^{3}/s. During the first decade of XXI century almost every two years the flood wave was bigger than the biggest one which occurs once in a fifty years. Huge flood waves were occurred in 2001, 2004, 2006, 2008 and 2010. The last flood in December 2010 had already reached the maximum value which occurs once in a hundred years (according to probability calculation (until and including) year of 2000)). Since the floods are directly and indirectly related to bank erosion these data should be included in bank erosion analysis because their analogy is proved, although there is no quantification of their correlation. Therefore, researches should be focused on causes of floods, and on reduction of bank erosion uncertainties. Many factors that influence the Kolubara River floods are already known. Firstly, there is a difference in flows in the upper and lower part of the Kolubara River basin. The drainage conditions in the upper part are more favorable. The area of hydrological profile Slovac is less than 1/3 of the whole basin, but it drains a half of all waters in the Kolubara River basin. Downstream hydrological profile Beli Brod encompasses a half of the basin, and its discharge is 3/4 of Kolubara's discharge. In the Donjokolubarski basin the drainage conditions are different, and the most significant factor is slope (the slope of the river flow and the slope of the river basin). The distance between Beli Brod and the Kolubara's confluence with the Sava River is 50 km and the altitude difference is 20 m. The present slope of 40 cm/km (0.4 ‰) is declining every year due to intensive sediment accumulation in the riverbed. Relating these processes with the shape of the river basin and rapid concentration of water downstream of the Beli Brod, it does not surprise that Kolubara River "ramp" over its alluvial plain. Moreover, in the last decade there is simultaneity of frequent rains of high intensity with extended duration and sudden snowmelt, and that is the reason for increased concern. Considering that rivers in the sub basins are mostly torrential, this concern is even more enhanced. Additionally, in this area rivers were diverted to bring the economic benefits. Because of all these reasons, the life in the coastal zone of the Kolubara River basin is gloomy but real with lot of uncertainties.

In order to prevent the frequent floods there are a several plans to deal with the actual situation in the area. Construction of several small accumulations on the Kolubara's tributaries is at its first phase, but there is no indication for solving the existing water problems. There is an idea to channelize the Kolubara's riverbed for sailing (i.e. for the transportation of lignite from the Kolubara mine), but it is still in the early phase of planning, although the initiative appeared long time ago. The height difference between the Kolubara's River confluence with Sava River and the location of lignite exploitation in the Kolubara mine basin is 23 m [12]. This height difference and the wideness of the Kolubara's riverbed would facilitate its riverbed training works, enabling cheaper lignite transportation from the Kolubara mine basin. Training works the Kolubara's riverbed and its preparation for lignite transportation could easily be carried out, so the invested means would be economically justified. Thus, the meandering flow would be straightened, and the strong bank erosion in the riverbed would be regulated which means that some factors of flooding would be eliminated.

CONCLUSION

Bank erosion, soil loss, sediment load deposition, changes in the river course, floods, landslides, soil and water pollution are the major environmental problems in the Kolubara River basin which could be aggravated by the land-use changes. The solutions for all mentioned environmental problems demand a complex analysis of the area characteristics and development of the strategy for solving the existing water problems in this area, but in the same time they have to provide necessary conditions for the further lignite exploitation. Some villages are located in the lower part of the Kolubara River basin, in the area which is planned for the expansion of the Kolubara mining basin, so it is an important factor for the future sustainable landscape planning.

Hydrological network of the Donjokolubarski basin is constantly changing due to natural factors and anthropogenic impacts. The damage which is done cannot be compensated, but even worse is the fact that no one feels responsible and that the population in this area is still left to the mercy of torrential river. Numerous calls for helping endangered people and goods were sent to the different addresses, but no one tried to help. Apparently, the problem goes beyond the "values" of a few villages and the state interest (lignite exploitation) has absolute priority, like in the case of neighboring Dubrava and unique sources of Obrenovac Municipality [34]. This situation lasted till the catastrophic floods in June 2010, when the shocking images of flood damage terrified the publicity, and problem could not be ignored anymore. As an attempt to repair the flood consequences, during 2010 two dikes were constructed with the length of 200 m in total. The first location was repaired for bridge protection, and the second one for household protection. The total cost of construction works was 100 000 euros. Since, the total length of all degraded river banks of first category is about 5 km; the economic profitability of this repairing method is questioned. It made sense in the initial phase of degradation, but now it goes beyond the reality of existing situation. It seems that, after the construction of two dikes, somebody tries to justify the negligence, because it is obvious that these two dykes are insufficient to solve the problem. In cases like this one, even not doing anything for protection of degraded areas represents a serious violation of principles of sustainable management of natural resources, actually that is an offence. The responsible for effects of the changes in the Kolubara River basin is still unknown, is it nature or man?

Making the constant pressure on state institutions through various appeals, indicating to unsustainability of current situation and stand by position of constant fear, this paper is one of many attempts to help the endangered population. In this context, the monitoring of the Kolubara River in the Donjokolubarski basin is a logical solution and our contribution, with particular results and recommendations, to fight for the basic human right to live without fear from hazards.

What kind of message can be sent to people living in this area and dealing with above mentioned problems? As they say, finding that the state does not protect them from the problems that come upon them, they give up farming the parcels of endangered area (along the river). The even more irrational, is the fact that they still pay taxes on the parcels, which does not exist anymore or they are significantly reduced, because the taxes calculation is made according to Cadaster from 1967! The estimation of all unnecessary loss of land, land values, personal losses of individuals and damage done to whole community is in the course. At the time when the personal status is far beyond collective responsibility due to difficult economic situation, this scientific approach is the only way to inspire the responsible ones in finding the solution.

This research could be the warning for the future anthropogenic activities on the river system since the new changes on the hydrological network were planned in this area. The four new mining fields should be opened, and if it happens, the hydrological network will be changed again and new problems will appear in the river basin.

ACKNOWLEDGEMENT

This paper was realized as a part of the projects "Studying climate change and its influence on the environment: impacts, adaptation and mitigation" (43007) and "The Democratic and National Capacities of Serbia's Institutions in the Process of International Integrations" (179009) financed by the Ministry of Education and Science of the Republic of Serbia within the framework of integrated and interdisciplinary research for the period 2011-2014. Translation and language correction was performed by Ljiljana Stanarevic.

REFERENCES

1.   S. Dragićević, M. Stepić, 2006Changes of the erosion intensity in the Ljig River basin- the influence of the antropogenic factor. Bull. Serbian Geogr. Soc. 85(2): 37-44 (in Serbian with English abstract)

2.   S. Dragićević, 2007Dominant Processes of Erosion in the Kolubara Basin. Faculty of Geography, Belgrade: Jantar groupe, 1245in Serbian

with summary in English).

3. S. Dragićević, I. Carević, S. Kostadinov, I. Novković, B. Abolmasov, B. Milojković, D. Simić, 2012Landslide susceptibility zonation in the Kolubara river basin (western Serbia)- analisys of input data. Carpathian Journal of Earth and Environmental Sciences 723747

4. S. Dragićević, I. Milevski, 2010Human Impact on the Landscape- Examples from Serbia and Macedonia. Advances in GeoEcology, 41Global Change- Challenges for soil management (Editor M. Zlatic), CATENA VERLAG GMBH, Germany. 298309

5. R. Tošić, 2006Soil erosion in the catchment Ukrina. Geographic Society of the Republic of Srpska, Banja Luka. Special issue, 13150in Serbian with summary in English)

6. V. Blanka, T. Kiss, 2011Effect of different water stages on bank erosion, case study of river Hernad, Hungary. Carpathian Journal of Earth and Environmental Sciences. 62101108

7. I. Milevski, 2011Factors, Forms, Assessment and Human Impact on Excess Erosion and Deposition in Upper Bregalnica Watershed (Republic of Macedonia). In: Human Impact on Landscape, Eds. S. Harnischmachter and D. Loczy. Zeitschrift für Geomorphologie, 55Suplementary 1Stuttgart, 7794

8. J. Chen, J. Z. Chen, M. Z. Tan, Z. T. Gong, 2002Soil degradation: a global problem endangering sustainable development. Journal of Geographical Sciences, 122243252

9. A. Goudie, 2006The human impact on the natural environment: past, present and future. Blackwell Publishing, USA, sixth edition, 357

10. L. Li, X. Lu, Z. Chen, 2007River channel change during the last 50 years in the middle Yangtze River, the Jianli reach. Geomorpholgy. 85185196

11. L. Denes, 2010Anthropogenic Geomorphology in Environmental Management. In: Anthropogenic Geomorphology- A Guide to Man-Made Landforms. Eds. József S., Lóránt D., Dénes L., Springer, 2538

12. S. Dragićević, N. Živković, V. Ducić, 2007Factors of flooding on the territory of the Obrenovac municipality. Collection of the papers, Faculty of Geography, Belgrade, 553954

13. M. Roksandić, 2012Causes and consequences of changes of hydrographic network in Donjokolubarski basin. Unpublished PhD thesis, University of Belgrade, Faculty of Geography, 197in Serbian with summary in English)

14. S. Dragićević, N. Živković, S. Kostadinov, 2008aChanges of hydrological

system in the lower course of the Kolubara river. In proceedings of the XXIV Conference of the Danubian countries on the hydrological forecasting and hydrological bases of water management, Bled, Slovenia.

15. M. Roksandić, S. Dragićević, N. Živković, S. Kostadinov, M. Zlatić, M. Martinović, 2011Bank erosion as a factor of soil loss and land use changes in the Kolubara river basin, Serbia. African journal of agricultural research, 63266046608DOI:AJAR11.736

16. S. Dragićević, 2002Sediment Load balance in the Kolubara basin. Faculty of Geography, Belgrade. 184in Serbian with summary in English)

17. S. Dragićević, M. Stepić, M. Karić, 2008bNatural potentials and degraded areas of Obrenovac municipality. Jantar groupe, Belgrade. 1180in Serbian with summary in English)

18. S. Dragićević, N. Živković, I. Novković, 2011Preparation of numerical and spatial data basis for the assessment of land and water diffuse pollution in the Kolubara River basin. Ministry of Environment, Mining and Spatial Planning, Environmental Protection Agency, Beograd (in Serbian)

19. J. Hooke, C. E. Redmond, 1989River-channel changes in England and Wales. Journal of Institution of Water and Environmental Management, 3328335

20. Large R G A, Petts E G1996Historical channel-floodplain dynamics along the River Trent, Implications for river rehabilitation. Applied Geography, 163191209

21. Q. Weng, 2002Land use change analysis in the Zhujiang Delta of China using satellite remote sensing, GIS and stochastic modeling. Journal of Environmental Management, 3273284

22. T. Kiss, K. Fiala, G. Sipos, 2008Alterations of channel parameters in response to river regulation works since 1840 on the Lower Tisza River (Hungary). Geomorphology, 9896110

23. R. C. De Rose, L. R. Basher, 2011Measurement of river bank and cliff erosion from sequential LIDAR and historical aerial photography. Geomorphology, 126132147

24. Cadastral register of the Municipality of Obrenovac,1967

25. D. Dukić, 1974Kolubara's regime and water management problems in its river basin. Bulletin of Serbian Academy of Science and Arts, Department of Science, 36Belgrade (in Serbian).

26. N. Surian, M. Rinaldi, 2003Morphological response to river engineering and management in alluvial channels in Italy. Geomorphology, 50307326

27. Youdeowei P O1997Bank collapse and erosion at the upper reaches of the Ekole creek in teh Niger delta area of Nigeria. Bulletin of the International Association of Engineering Geology 55167172

28. ***Diverting of Kolubara River: http://www.neshvyl.com/doc/prica_o_ kolubari.pdf www. rbkolubara.rs

29. J. Hooke, 1995River channel adjustment to meander cutoffs on the River Bollin and River Dane, northwest England. Geomorphology, 14235253

30. R. Pejanović, Z. Njegovan, 2009Actual problems of Serbian agriculture and villages. Industry. 37(1): 87-99 (in Serbian)

31. Republic Hydro Meteorological Service of Serbia. Values of precipitation, water discharge, sediment load concentration for Kolubara river basin (1961-2005

32. D. Chen, J. G. Duan, 2006Modeling with adjustment in meandring channels. Journal of hydrology, 3215976

33. S. Dragićević, S. Nenadović, B. Jovanović, M. Milanović, I. Novković, D. Pavić, M. Lješević, 2010Degradation of Topciderska River water quality (Belgrade). Carpathian Journal of Earth and Environmental Sciences, 52177184

34. N. Zivkovic, S. Dragicevic, I. Brceski, R. Ristic, I. Novkovic, S. Jovanovic, M. Djokic, S. Simic, 2012Groundwater Quality Degradation in Obrenovac Municipality, Serbia. In: K. Voudouris and D. Voutsa (Eds.), Water Quality Monitoring and Assessment, Rijeka: InTech. 283300

35. W. De Vries, D. J. Bakker, 1998Manual for calculating critical loads of heavy metals for terestial ecosystems. Guidelines for critical limits, calculation methods and input data. Wageningen, DLO Winand Staring Centre. Report 166. 144

36. OSPAR/ICES2004Workshop on the evaluation and update of background reference concentrations (B/RCs) and ecotoxicological assessment criteria (EACs) and how these assessment tools should be used in assessing contaminants in water, sediment and biota, 913February 2004. The Hague, Final report, OSPAR Commission, 1-90442-652-2

Chapter 8

ENVIRONMENTAL LAND USE AND THE ECOLOGICAL FOOTPRINT OF HIGHER LEARNING

Seth Appiah-Opoku[1] and Crystal Taylor[2]

[1]Geography Department, University of Alabama, Tuscaloosa, USA
[2]Florida State University, USA

INTRODUCTION

The lifestyles of individuals, groups, or nations can be measured by utilizing an accounting tool known as ecological footprint. Ecological footprint refers to the productive land needed to support a given population. As discussed by Wackernagel and Rees (1996), "The ecological footprint concept is simple, yet potentially comprehensive: it accounts for the flows of energy and matter to and from any defined economy and converts these into the corresponding land/water are required from nature to support these flows" (p. 3). A concept known as "overshoot" occurs if demands by humans exceed the supply of a given biologically productive area (Turner et al., 2006). Thus, a larger ecological footprint indicates a less sustainable society.

Research on ecological footprint literature links together the concepts of footprint size and economic development. In other words, footprints represent population size and consumption levels (Wackernagel & Rees, 1996). Furthermore, more-developed countries contain market economies that consume greater levels of natural resources, and environmental degradation is largely driven by the growth and intensification of market economies (Jorgenson, as cited in Jorgenson & Burns, 2006). For example, Americans when compared to the rest of the world exhibit a large ecological footprint due to an intensely consumption-oriented lifestyle. The average ecological footprint for an American is 23.68 acres as compared to the world's average of 5.53 acres (Global Footprint Network, 2003). Further research suggests an economical discrepancy between those who possess large ecological footprints and those who possess small ecological footprints. Wackernagel et al. (2003) found that those contributing most to climate change through their energy intensive lifestyles will most likely be less affected by, and better shielded from, the outfalls of climate change than poor people living on marginal land or in underserved urban conditions.

Though ecological footprint can be used as a useful tool to help measure sustainability, some scientists have criticized ecological footprint calculations for oversimplifying ecosystem processes to numerical values. Assumptions may not be valid as the ecological footprint arbitrarily assumes both zero greenhouse gas emissions, which may not be optimal, and national boundaries, which makes extrapolating from the average ecological footprint problematic (Fiala, 2008). Despite these criticisms, the ecological footprint calculation can serve as a heuristic tool for designing and implementing plans for today as well as for tomorrow. Moreover, plans that take environmental calculations into consideration will have a far greater potential of keeping the Earth as a stakeholder in the planning process than those plans without such calculations.

Colleges and Universities across the world serve as incubators for tomorrow's leaders. In essence, they leave an educational imprint on individuals in an effort to educate and facilitate the development of tomorrow's leaders. These institutions serve as the setting where ideas can take form and this is where ideas can be implemented in a semi contained setting as part of the larger community. Though it is well established that educational institutions leave their imprints on innovative minds, this chapter introduces the idea that institutions of higher learning also leave ecological footprints on the landscape. Universities provide support to environmental issues through policies, programs, and research. The idea of greening campuses has become so popular that the Princeton Review has posted a Green Rating Honor Roll to document the top schools that provide a healthy and sustainable quality of life for the students, environmentally-minded and educational preparations for the future workforce, and environmentally responsible school policies for all to follow (Princeton Review, 2008).

Thinking green has been a hot topic among US Colleges in recent years. To think green is to incorporate environmental impacts into decision-making activities that affect daily lifestyles. The impact that a society imposes on the environment holds importance, as it is a key issue of sustainability. Sustainability refers to the dilemma of how to "meet the needs of the present without compromising the ability of future generations to meet their own needs" (Wackernagel & Rees, 1996, p. 33). Fortunately, one place where sustainable initiatives have spread is on campuses throughout America. Universities have provided support to environmental issues through policies, programs, and research.

The Princeton Review has posted a Green Rating Honor Roll to document the top schools that provide a healthy and sustainable quality of life for the students, environmentally-minded and educational preparations for the future workforce, and environmentally responsible school policies for all to follow (Princeton Review, 2008). The Princeton Review ranked the following eleven colleges throughout the United States as receiving a green rating of ninety-nine points.

- Arizona State University, Tempe
- Bates College
- College of the Atlantic
- Emory University
- Georgia Institute of Technology
- Harvard University
- State University of New York at Binghamton
- University of New Hampshire
- University of Oregon
- University of Washington
- Yale University

Although all the above-listed universities have displayed an extraordinary commitment to green initiatives, Harvard University located in Cambridge, MA; Emory University located near Atlanta, GA; and Bates College in Lewiston, ME, were chosen for closer examination in part due to the accessibility of online information concerning green programs as well as in respect to their diverse financial strategies for integrating sustainable principles. An inventory was performed encompassing a list of similarities and differences concerning green initiatives and strategies. Moreover, this inventory can serve as a framework for other colleges to follow in the future. It is in this context that we discuss the current consumption and environmental awareness levels associated with the use of water and energy resources for dormitory students on The University of Alabama's campus.

The University of Alabama is in the preliminary stages of moving toward a more sustainable campus. Currently, it is difficult to track environmentally friendly progress on campus, as no study has been previously performed to establish where The University of Alabama is concerning environmental initiatives. Thus, if a snapshot of the University were established to include both consumption and environmental awareness levels, then those findings would serve as a benchmark from which the implementation of green strategies may be evaluated in terms of effectiveness. Accordingly, this research documented the environmental awareness and consumption levels of dormitory students concerning energy and water resources on The University of Alabama's campus. Moreover, findings were gathered from the dormitories Ridgecrest East and Lakeside East. The goal was to measure the current ecological footprint of dormitory students on The University of Alabama campus. Specific research objectives were to (a) determine the current state of students' environmental awareness, and (b) determine the current consumption levels in terms of electricity and water usage for specific dormitories on campus.

RESEARCH METHODS

A case study approach was utilized during this research. According to Theodorson and Theodorson (as cited in Punch, 1998) a case study is defined as "a method of studying social phenomena through the thorough analysis of an individual case. Described simply, a case study provides a snapshot of particular social phenomena (Hakim, 1987, p. 61). Thus, the case study approach allows for in-depth research on specific populations, such as the dormitory students that will serve as the focus for this research. This approach also permits the researcher to evaluate subjects in a naturalistic setting as well as conduct research from a wide array of methods such as interviews, observations, numerical data, and questionnaires (Punch, 1998, p. 153). Suitably, interviews, observations, surveys, and data analysis are the primary methods utilized in this research. Even as the case study approach proves to be a viable research tool, a limitation is the inability of the researcher to derive generalizations from specific instances (Punch, 1998, p. 155). In light of this accusation, it is of importance to note that the case study approach warrants merit as this research requires an in-depth inquiry into a particular situation that has yet to be documented.

As mentioned previously, the focus of this research is centered around dormitory students residing on The University of Alabama campus located in Tuscaloosa, Alabama. In the fall of 2008, The University of Alabama reached a record enrollment of 27,052 students (Andreen, 2008). Of the 27,052 students approximately 7,000 students are housed on campus (E. Russell, e-mail, February 24, 2009).[1] -Therefore, on-campus residency accounts for approximately 26% of the student population as illustrated by Figure 1.

For this study two dormitories were chosen for sampling. The selection was done by methods of random sampling. Random sampling allowed every dormitory to have an equal opportunity of being selected. The process entailed writing down the names of all the possible dormitories on campus on individual slips of paper. The dormitory names were mixed up and then drawn out of a hat. The dormitories Lakeside East and Ridgecrest East were selected for an analysis of energy and water usage records. The coed student populations housed within Lakeside East and Ridgecrest East are 238 and 316 students, respectively.

UA Student Housing

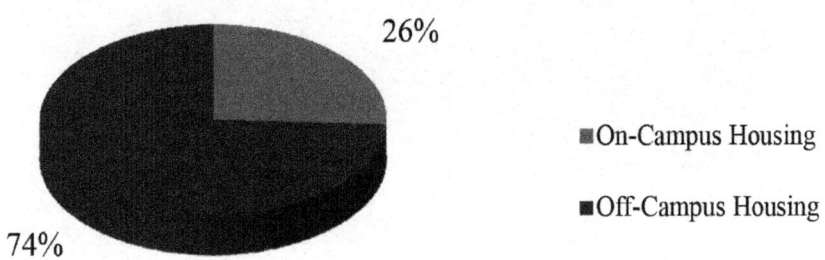

Figure 1: UA student housing.

SURVEY AND DATA ANALYSIS

The University of Alabama's Department of Energy Management aided in providing energy and water consumption records concerning the Ridgecrest East and Lakeside East dormitories. A content analysis of the records was performed to determine applicable

Figure 2: Lakeside East Residential Hall.

Figure 3: Ridgecrest East Residential Hall.

themes and patterns. Additionally, the records assisted with the calculation of the ecological footprint analysis of energy and water usage in dormitories on campus. The energy records acquired reported monthly electrical and natural gas usage figures for the two dormitories from 2007 and 2008. Due to some technical problems with the water meters, only the last five months of 2008 were available for analysis. However, water usage assumptions were derived for the entire year of 2008. Ecological footprint calculations were projected from estimates of the average water usage in 2008 and from the actual natural gas and electricity usage figures from 2007 and 2008. Even from water approximations, the derived ecological footprint has the ability to serve as a benchmark that can be utilized in future research. During the analysis of water and energy records, the data concerning the population rates for Ridgecrest East and Lakeside East during 2007 were unfortunately unattainable; consequently, the 2008 population numbers were substituted. In addition to the analysis of energy records, an interview with the Director of Energy Management was conducted in an effort to get a proper vision of the campus in terms of resource management.

CALCULATING THE ECOLOGICAL FOOTPRINT

Data from the Department of Energy Management were utilized in the ecological footprint calculation. The following identifies the process for calculating ecological footprints:

- Estimate the average population size.
- Estimate the average annual consumption for a particular item.
- Estimate the land area appropriated per capita for the production of items consumed.
- Estimate the ecological footprint of the average person for all items consumed.
- Multiply the population by the per capita footprint.

The ecological footprint calculation was utilized to determine land use requirements associated with the consumption of resources. The calculation was performed utilizing water, electric, and natural gas records. All the records used in this study were obtained from the Department of Energy Management.

Water

Water usage records were acquired pertaining to Ridgecrest East and Lakeside East Residential Halls from August to December 2008. Due to some technical problems with the water meters, accurate water usage readings prior to August 2008 were unattainable. The trend for water usage at Ridgecrest East showed little variation during the months of August, September, and October as consumption ranged from approximately 72,000 to 80,000 cubic feet or approximately 538,000 gallons to 599,000 gallons. Usage dropped slightly during November followed by a dramatic decrease in December. Lakeside East Residential Hall demonstrated more drastic trends than Ridgecrest East as usage in August peaked at nearly 140,000 cubic feet followed by a marked decline in September as Lakeside levels dropped around 40%. A slight increase occurred during the month of October. In November and December consumption decreased drastically as water usage dipped below Ridgecrest levels.

Figure 11 details water usage in cubic feet consumed. Figure 12 depicts the steps we were utilized to calculate the ecological footprint of water resources consumed in the dormitories Lakeside East and Ridgecrest East. First, the populations of Lakeside East and Ridgecrest East were established. As mentioned previously, 238 students reside within Lakeside East, whereas 316 students live in Ridgecrest East. A full twelve months of records were unavailable, so estimations were used to approximate the yearly water consumption levels within the dormitories. The 2008 yearly estimations for each building were derived from taking the average amount of water used during the five months and then multiplying that average by twelve months. For Ridgecrest East the figure 746,616 cubic feet was used as the 2008 water usage estimate, while the figure 911,496 cubic feet was used for Lakeside East.

Thus, an ecological footprint calculation concerning water resources can be derived by utilizing the water consumption estimates for the two residential halls as indicated above. Initially, the amount of water consumed in cubic feet per dormitory student must be established. The number was calculated by dividing the total water estimates for each

Source: University of Alabama Department of Energy Management

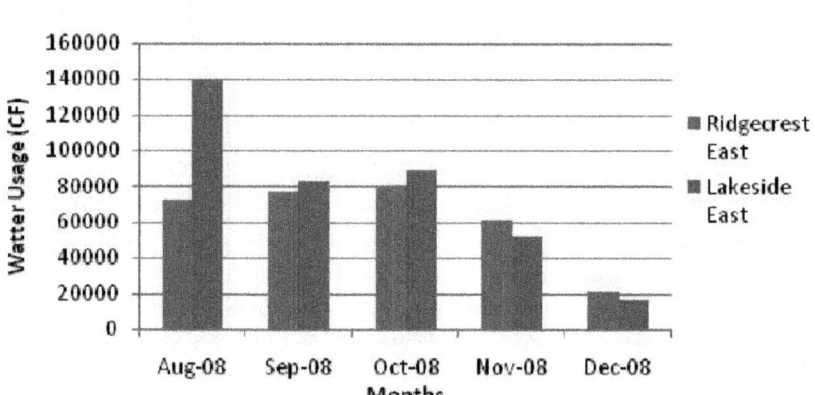

2008 Water Usage for Ridgecrest East & Lakeside East

Figure 4: Water Usage for Ridgecrest East and Lakeside East.

Step 1	Step 2	Step 3	Step 4	Step 5
Identify the population size of the dormitories	Obtain water records	Determine land area requirements for water resources	Estimate the Ecological Footprint for the average person	Multiply the per capita footprint by the total population on campus

Figure 5: Ecological footprint procedure for water.

dormitory by the subsequent student populations residing in each residential hall. Consequently, the average amount of water consumed per student for Lakeside East was 3,830 cubic feet (28,649 gallons) and 2,363 cubic feet (17,674 gallons) for Ridgecrest East.

To obtain a real-world comparison, consumption figures of the individual dormitory student are listed in gallons as well as cubic feet. The individual usage levels can further be broken down into daily usage figures by dividing by 365 to represent the approximate number of days in a year. As a result the daily consumption level for an individual residing in Lakeside East was 10.49 cubic feet or 78.49 gallons and 6.47 cubic feet or 48.42 gallons for those in Ridgecrest East. Daily usage figures are useful as they can be easily compared to the national average of the average American. According to the Environmental Protection Agency (2003), the average American consumes 90 gallons of water daily in the home, as compared to the average European consuming 53 gallons daily, and the typical Sub-Saharan African citizen consuming only 3-5 gallons per day.

After establishing the consumption levels for water resources, it was necessary to determine the amount of land required for the utilization of water resources. Thus, water resources were converted to cubic meters by multiplying by 0.0283 and then divided by 1,500 $m^3/ha/yr$ to accommodate the amount of forested land needed to accommodate the water consumed (Anundson et al., 2001, p. 26). The result was equivalent to

0.0723 hectares (0.1785 acres) per dormitory student in Lakeside East and 0.0446 hectares (0.1101 acres) per dormitory student in Ridgecrest East

Table 1: Ecological Footprint for Water 200

Ecological Footprint for Water 2008	Lakeside East	Ridgecrest East
Total Water Usage 2008 (cubic ft)	911,496	746,616

Water Usage per Month (cubic ft)	75,958	62,218
Water per Student in 2008 (cubic ft)	3,830	2,363
Total Land Area in Hectares per Dormitory Student	0.0723	0.0446
Total Land Area in Acres per Dormitory Student	0.1785	0.1101

It is germane to keep in mind that all of these figures, concerning hectares/acreage required, only apply to the land required concerning water resources utilized during the consumption of housing. Accordingly, "the ecological footprint concept is based on the idea that for every item of material or energy consumption, a certain amount of land in one or more ecosystem categories is required to provide the consumption-related resource flows and waste sinks" (Wackernagel & Rees, 1996, p. 63). Thus, a complete ecological footprint calculation encompasses many different goods and services as this study looks specifically at water and energy resources associated with housing needs of dormitory students on The University of Alabama's campus.

Electricity

In addition to supplying the water records, as indicated in the findings in the previous section, the Department of Energy Management also provided electric and natural gas records for use in this research. To assist with the analysis of Lakeside East and Ridgecrest East Residential Communities, complete electrical and natural gas records were gathered from January 2007 to December 2008. Energy consumptions records from both 2007 and 2008 show a general trend of Lakeside East utilizing slightly less electricity per month with the exception of a peak on September 2007. During September 2007, Lakeside East Residential Hall experienced a spike in usage as

315,007 kilowatt hours (kWh) were consumed. This consumption stands-out on the electrical records as neither Lakeside East nor Ridgecrest East demonstrated another usage level over 140,000 kilowatt hours during the two-year span.

Despite the September peak for Lakeside East, electricity usage throughout the 2007 year remained somewhat consistent as January through March accounted for a range of approximately 50,000 to 65,000 kWh. April to May experienced a slight increase with consumption hovering near 80,000 kWh. June to July numbers were barely below 70,000 kWh, while August numbers increased back up to nearly 80,000 kWh. October boasted the second highest usage for 2007 at 87,151 kWh. Finally, during the months of November and December consumption ranged from 65,000 to 55,000 kWh. Interestingly, even as Lakeside East consistently consumed less power per month during 2007 with the exception of the September spike, the total 2007 energy consumption figures for Lakeside East (1,067,609 kWh) were slightly higher than Ridgecrest East (1,066,400 kWh).

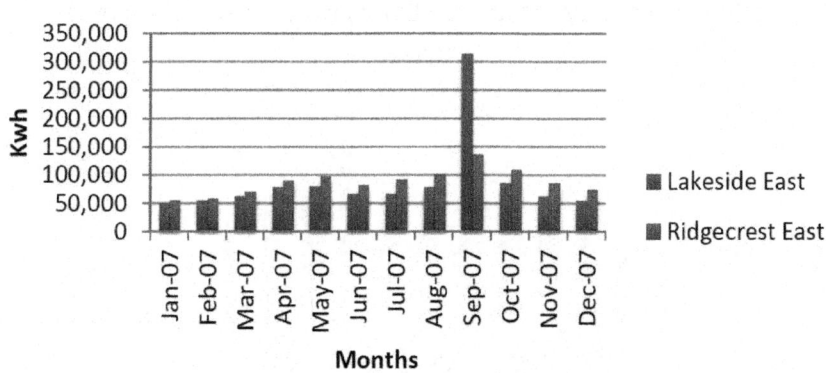

Source: University of Alabama Department of Energy Management

Figure 6: Electricity Usage for Lakeside East and Ridgecrest East.

As mentioned previously, Ridgecrest East has in general consumed a higher amount of electricity in terms of kilowatt hours per month during 2007 when compared to Lakeside East. Those higher consumption rates for Ridgecrest East are indicated as the following approximated percentages above Lakeside East's usage levels: January was 11% higher, February displayed an 8% increase, March had an 11% increase, April saw a 16% increase, May's increase jumped up 22%, June displayed a 24% increase, July had a 35% increase, October displayed a 27% rise, November increased to 35%, and finally December had a 36% increase over Lakeside East's consumption levels. Electricity consumption for Ridgecrest East during September 2007 was only about 44% of what Lakeside East consumed.

During 2008, Lakeside East consumed less total electricity each month than Ridgecrest East. Moreover, when the total consumption figures of 2008 for both dormitories are compared to the 2007 fiscal year, together the buildings show an overall decrease in electrical usage. Lakeside East displayed the following monthly consumption during 2008 recorded in kilowatt hours: January was 39,628 kWh; followed by February with 62,320 kWh; March consumed 58,206 kWh; April used 65,469 kWh; May was 68,613 kWh; June was recorded at 59,222 kWh; July had 62,597 kWh; August consumed 72,264 kWh; September was recorded at 108,040 kWh; October used 81,022 kWh; November had 60,623 kWh of usage; and finally during December 57,632 kWh were utilized. Similar to the methodology utilized to calculate the ecological footprint concerning water resources, Figure 15 depicts the ecological footprint procedure from which the electrical impact of students was derived.

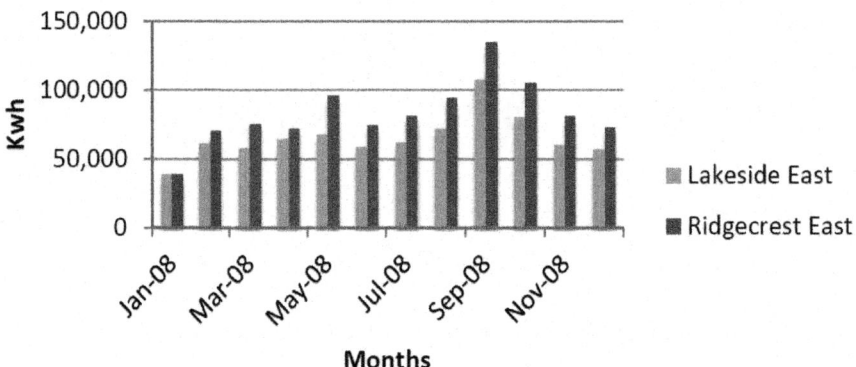

Source: University of Alabama Department of Energy Management

Figure 7: Electricity Usage for Lakeside East and Ridgecrest East.

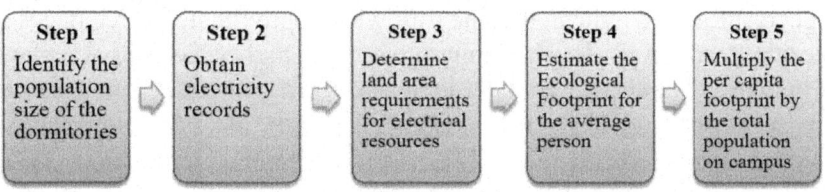

Figure 8: Ecological footprint procedure for electricity

For a more in-depth analysis of electrical usage for the two dormitories, the amount of energy utilized by each dormitory student for the year was calculated as the total electricity consumption numbers were divided by the amount of students residing within each dormitory. This accounted for the amount of electricity utilized per student to be 4,486 kWh at Lakeside East and 3,375 kWh at Ridgecrest East. It is important to note that even though the energy consumption numbers showed little variation during the 2007 fiscal year, the higher population numbers within Ridgecrest East resulted in energy usage per student that was considerably less than those found at Lakeside East. Just as the 2007 electricity records were broken down for analysis, the 2008 electricity records were evaluated for individual usage levels.

To acquire the electricity consumed per dormitory student during 2008, the electrical totals were divided by the amount of the respective residential populations. Thus, the average student consumed 3,343 kWh within Lakeside East and 3,177 kWh for Ridgecrest East. To relate student electricity consumption rates to a real-world example the 2007 and 2008 figures were broken into monthly averages. The 2007 monthly rates per dormitory student were calculated to be approximately 374 kWh for Lakeside East and approximately 281 kWh for Ridgecrest East. For 2008 the monthly averages were approximately 279 kWh for Lakeside East and approximately 265 kWh for Ridgecrest East. According to the Energy Information Administration (2007), the average Alabama household consumes 1,305 kWh per month.

After the consumption levels were successfully calculated for electrical resources, the amount of land could be determined for the usage of electrical resources. To accommodate the carbon emissions from the utilization of electricity the rate of 169 m^2 of forest for every 100 kWh of electricity was used for the following ecological footprint calculations (Anundson et al., 2001, p.11). Thus, the individual amount of electricity per dormitory student was first divided by 100 kWh and then multiplied by 169 m^2. Accordingly during 2007 for Lakeside East, the amount of land needed per dormitory student was 7,581 m^2 (0.758 hectares or 1.873 acres) and for Ridgecrest East 5,703 m^2 (0.570 hectares or 1.409 acres). During 2008, the amount of forested land area necessary per student amounted to 5,650 m2 (0.565 hectares or 1.396 acres) for Lakeside East and 5,369 m^2 (0.537 or 1.327 acres) for Ridgecrest East. In Table 7, meters squared were converted to hectares by dividing by 10,000. Additionally, hectares were converted by multiplying by 2.471.

Table 2: Ecological Footprint for Electricity 2007 and 2008

Ecological Footprint for Electricity 2007	Lakeside East	Ridgecrest East

Total Electricity 2007 (kWh)	1,067,609	1,066,400
Electricity per Student in 2007 (kWh)	4,486	3,375
2007 Total Land (m)^2 per dormitory student	7,581	5,703
2007 Total Land in Hectares per dormitory student	0.758	0.570
2007 Total Land in Acres per dormitory student	1.873	1.409
Ecological Footprint for Electricity 2008	**Lakeside East**	**Ridgecrest East**
Total Electricity 2008 (kWh)	795,636	1,004,000
Electricity per Student in 2008 (kWh)	3,343	3,177
2008 Total Land (m)^2 per dormitory student	5,650	5,369
2008 Total Land in Hectares per dormitory student	0.565	0.537
2008 Total Land in Acres per dormitory student	1.396	1.327

As a reminder, it is important to note that all the ecological footprint analysis that has been mentioned in this section pertains only to the electrical energy consumption as related to housing concerns. In reality electricity consumed for housing is only one area of a person's life where electricity is utilized. Therefore, the electrical usage and subsequent land area may in fact be larger than the estimates listed above. In general, ecological footprint calculations encompass a variety of goods and services associated with a person's lifestyle. This research looked specifically at water and energy usage of the footprint equation as related to housing needs.

In addition, each student's consumption of natural gas was calculated in the same way. Thereafter, each students total land area requirement at Lakeside East was calculated as follows: 0.179 acres for water resources in 2008, 1.873 acres for electricity in 2007, 1.396 acres for electricity in 2008, 0.170 acres for natural gas in 2007, and 0.177 acres for natural gas in 2008. Furthermore, Ridgecrest East's numbers were 0.110 acres for water in 2008, 1.409 acres for electricity in 2007, 1.327 acres for electricity in 2008, 0.142 acres for natural gas in 2007, and 0.140 acres for natural gas in 2008. Thus, if the entire student population that resides on-campus of approximately 7,000 individuals adopted the consumption habits of either Lakesides East or Ridgecrest East residents, then the land acreage as illustrated in Table 7 would have been needed.

When evaluating these figures it is important to understand that Lakeside East and Ridgecrest East are both relatively new buildings found on The University of Alabama's campus. As this study represents a sample of consumption levels taken from the new and therefore more efficiently constructed dormitories, the land requirement estimations for the students living on-campus are likely to be a best-case scenario. Overall, from the ecological footprint calculations utilized, Ridgecrest East displayed a lower environmental impact or land requirement than Lakeside East for water, electricity, and natural gas.

Additionally, land requirements decreased for electricity needs for both dormitories from 2007 to 2008. On the other hand, during the two year-span the land requirements for natural gas showed only a slight decrease for Ridgecrest East while Lakeside East showed an increase in demand. Acreage for water resources were not compared from 2007 to 2008 as the required data were unattainable.

Table 3: Ecological Footprint for the On-Campus Population

Ecological Footprint: Land Requirements in Acres for the Dormitory Student Population	Lakeside East	Ridgecrest East
From 2008 Water Consumed	1,253	770
From 2007 Electricity Consumed	13,111	9,863
From 2008 Electricity Consumed	9,772	9,289
From 2007 Natural Gas Consumed	1,190	994
From 2008 Natural Gas Consumed	1,239	980

CONCLUSION AND POLICY IMPLICATIONS

Although much progress has been made in recent years there is more that The University of Alabama can do in support of sustainable practices, as exemplified by green universities across the country. The first step toward becoming a green campus merely entails setting the goal of wanting to be more sustainable. The President of University of Alabama's message to the student body during fall of 2008 was the initial step required to set the tone for the campus. Now that a goal has been set, a subsequent plan will need to be developed. Objectives will need to be established in order to facilitate progress toward the end goal.

Before any other steps of the plan can be formulated lest carried out, it is essential to stop and take an inventory. The inventory determines where the campus is now so that progress may be more accurately measured. Thus, this research has served as a snapshot of where the campus currently is, during the academic semesters of fall of 2008 to early spring of 2009 in terms of sustainability. The snapshot is a useful tool as it was used to compare The University of Alabama to the top green schools. These prestigious universities were utilized in this analysis to serve as the pinnacle of where The University of Alabama may strive to be concerning environmental initiatives.

Taking the other schools analyzed in this research into consideration, our first recommendation is to formulate an official environmental plan that involves a variety of stakeholders in the planning process. This initiative needs the involvement of students, faculty, staff, alumni, investors, and the community as a whole. During the planning process, objectives must be set that are measurable as well as quantifiable to the overall goal of the plan. If these objectives are to serve as serve as milestones towards the goal of sustainability. Ecological footprint calculations as used in this study will be beneficial for monitoring progress towards this goal.

Our second recommendation is to strive to establish a recognizable environmental office on campus supported by a full-time staff. This ensures availability of knowledgeable staff to assist with inquiries from environmentally-aware students and community members as well as to address sustainability issues in accordance with the campus's environmental plan. According to data gathered on sustainable universities by the Sustainable Endowments Institute (2009), a considerable number of schools have recognized the need for full-time campus sustainability administrators. Currently, 56 percent report having dedicated sustainability staff.

We also recommend the incorporation of green building elements within residential student housing just. Generally speaking, universities are long-term owners of institutions. Hence, looking at the cost of operation over the period of a product's life cycle will help them accept some of the additional costs associated with green building methods. According to Moskow (2008), "Sustainable developments are more cost-effective in the long term and, therefore, ultimately, more valuable" (p.xv). This is especially true as the price of resources such as electricity and natural gas continue to rise. Additionally, green buildings have been noted to promote a healthy, productive work environment that would benefit the welfare and academic status of The University of Alabama.

Fortunately, The University of Alabama has already begun incorporating some green features in buildings such as low flow toilets, low flow faucets, low flow showerheads as well as plans for lighting controls and high efficiency hoods for new projects. Though those efforts are commendable, our recommendation is to use Bates College as an example to strive toward concerning green buildings. Due to cost restrictions, Bates College has not filed for the proper LEED certification for their structures. Despite not having filed, Bates College has used the LEED criteria as a standard in which to construct LEED equivalent buildings. Furthermore, green is marketable and green building designs are a good way to promote The University of Alabama's image.

Our final recommendation is education. Additional educational opportunities may in fact reduce the environmental impact of the University. Due to the fact that the role of academic institutions is to educate and facilitate in the development of tomorrow's leaders, this is a prime environment within which to integrate green technologies. Leaders that are unable to recognize the mismanagement of resources will be incapable of solving environmental problems. If environmentally friendly strategies are to be incorporated into future policies, then exposure to sustainable education is essential.

An expansion of research concerning ecological footprint analysis would be beneficial in an effort to determine the environmental impact of the campus. Though food and recycling strategies were only briefly discussed in this study, a more in-depth analysis may be needed to evaluate whether or not the University should try to promote locally or organically supplied food in the cafeterias and whether or not to participate in the *RecycleMania* competition. Additionally as only dormitory students were analyzed in this study, more sample groups could be evaluated and include both on-campus and off-campus students. Studies on climate change, transportation issues, student led initiatives, and a plethora of other opportunities exist for exploration.

In conclusion, we wish to emphasize that if places of higher learning are able to lessen their ecological footprints, they would ultimately have a greater positive impact on humanity and the dwindling resources of the World.

REFERENCES

1. American College & University Presidents Climate Commitment.2008About Environmental Land Use and the Ecological Footprint of Higher Learningfrom http://www. presidentsclimatecommitment.org/html/about.php

2. C. Andreen, 2008UA Enrollment Reaches Record 27,052 Students; Freshman Class Tops 5,000.Retrieved February 26, 2009, fromhttp:// uanews.ua.edu/anewssep08/enrollment091608.htm

3. B. Anundson, J. Crooks, A. Fletcher, M. Frank, et al.2001A Study of the Ecological Footprint of Allegheny College. Unpublished manuscript.

4. 2008, November 20N. Barrella, (200, 20. November, competition. In, seeks. Harvard, "green. to, dorms. up", Harvard seeks to "green up"

dorms. Harvard Law Record. Retrieved December 3, 2008, fromhttp:// media.www.hlrecord.org/media/storage/paper609/news/2008/11/20/ News/In.Competition.Harvard.Seeks.To.green.Up.Dorms-3554348. shtml

5.   Bates College.2008Student Housing at 280 College Street. Retrieved December 2, 2008, from http://www.bates.edu/x175547.xml

6.   2008, September 10B. Bralley, (200, 10. September, U. A. Construction, Green. Goes, White. Crimson, September. 1. Retrieved, 200, fromhttp:// www.cw.ua.edu/ua

7.   2009, January 7K. Bursch, (200, 7. January, U. A. Starts, Green. New, Crimson. Campaign, White, Retrieved January 15, 2009, fromhttp:// www.cw.ua.edu/ua_starts_newgreen_campaign

8.   Emory University.2008aEmory Sustainable Initiative: History. Retrieved December 11, 2008, fromhttp://sustainability.emory.edu/page/1015/ History

9.   Emory University.2008bEmory Sustainable Initiative: Sustainable Food. Retrieved December 11, 2008, fromhttp://sustainability.emory.edu/ page/1008/Sustainable-Food

10.   J. Enck, S. Turner, 2003ASHRAE Green Guide: An ASHRAE Publication Addressing Matters of Interest to Those Involved in Green or Sustainable Design of Buildings. Atlanta: American Society of Heating, Refrigerating and Air-Conditioning Engineers, Inc.

11.   Energy Information Administration.2001Natural Gas Consumption and Expeditures in U.S. Households by End Uses and Census Region, 2001. Retrieved March 3, 2009, from http://www.eia.doe.gov/emeu/recs/ byfuels/2001/byfuel_ng.pdf

12.   Energy Information Administration.2007U.S. Average Monthly Bill by Sector, Census Division, and State 2007. Retrieved March 3, 2009, fromhttp://www.eia.doe.gov/cneaf/electricity/esr/table5.html

13.   Environmental Protection Agency.2003Water on tap: What you need to know.Retreived March 25, 2009, fromhttp://www.epa.gov/safewater/ wot/pdfs/book_waterontap_full.pdf

14.   Environmental Protection Agency.2008Wastes- Non-Hazardous Wastes. Retrieved October 12, 2008, fromhttp://www.epa.gov/epawaste/nonhaz/ index.htm

15.   Environmental Protection Agency.2009aBrownfields and Land Revitalization. Retrieved March 19, 2009, from http://www.epa.gov/ brownfields/

16. Environmental Protection Agency.2009bWastes- Resource Conservation-Reduce, Reuse, Recycle. Retrieved April 16, 2009, fromhttp://www.epa.gov/osw/conserve/rrr/reduce.htm

17. N. Fiala, 2008Measuring sustainability: Why the ecological footprint is bad economic and bad environmental science. Ecological Economics. Retrieved October 11, 2008, from ScienceDirect database.

18. T. Glavinich, 2008Contractor's Guide to Green Building Construction: Management, Project Delivery, Documentation, and Risk Reduction. New Jersey: John Wiley & Sons, Inc.

19. Global Footprint Network.2003National Footprints.Global Footprint Network. Retrieved October 1, 2008, fromhttp://www.footprintnetwork.org/gfn_sub.php?content=national_footprints

20. "Global Warming". (n.d.).New York Times. Retrieved October 11,2008fromhttp://topics.nytimes.com/top/news/science/topics/globalwarming/index.html

21. "Graduate Green Living Program Enters Its Second Year".2007, SpringHarvard Green Campus Newsletter. Retrieved December 4, 2008, fromhttp://www.greencampus.harvard.edu/newsletter/archives/2007/05/graduate_green.php

22. 2008, September 19K. Gray, (200, 19. September, Emory Freshman Live 'Green' in New Housing. [News Release]. Retrieved December 3, 2008, from http://www.emory.edu/home/news/releases/2008/09/green-dorms-open.html

23. C. Hakim, 1987Environmental Land Use and the Ecological Footprint of Higher LearningLondon: Allen & Unwin.

24. 2005, March 31B. Handwerk, (200, 31, Earth's Health in Sharp Decline, Massive Study Finds. National Geographic. Retrieved October 10, 2008, from http://news.nationalgeographic.com/news/2005/03/0331_050330_unenvironment.html

25. C. Harper, 2008Environment and Society: Human Perspectives on Environmental 4thed.). New Jersey: Pearson Education, Inc.

26. Harvard University.2008Green Campus Loan Fund: Harvard Office for Sustainability. Retrieved December 11, 2008, fromhttp://www.greencampus.harvard.edu/gclf/

27. 2008, November 13C. Ireland, (200, 13. November, Living in the green zone at "Rock Hall". Harvard University Gazette Online. Retrieved December 3, 2008, from http://www.news.harvard.edu/gazette/2008/11.13/11 -rockefeller.html

28. A. Jorgenson, T. Burns, 2006Environmental Land Use and the Ecological Footprint of Higher LearningSocial Science ResearchRetrieved October 11, 2008, from Science Direct database.

29. J. Kitzes, A. Galli, M. Bagliani, J. Barrett, G. Dige, Ede, S. , et al.2008Environmental Land Use and the Ecological Footprint of Higher LearningEcological EconomicsRetrieved October 11, 2008, from ScienceDirect database.

30. B. Kobet, S. Lee, C. Mondor, 1999Green Buildings: Environmental Land Use and the Ecological Footprint of Higher Learningection.

31. T. Leopard, 2008SGA Memo: University Efforts in Regard to Sustainability. [Memo]. University of Alabama.

32. 2007, AutumnM. Loftus, (200, Autumn, Emory sprouts new green residence halls. Emory Magazine. Retrieved December 3, 2008, fromhttp://www.emory.edu/EMORY_MAGAZINE/2007/autumn/halls. html

33. 2002, January 10H. Mayell, (200, 10. January, Green Group Gives Earth Failing Report Card. National Geographic. Retrieved on October 11, 2008, from http://news.nationalgeographic.com/news/2002/01/0110_02 0110worldwatch.html

34. K. Moskow, 2008Environmental Land Use and the Ecological Footprint of Higher LearningNew York: McGraw-Hill

35. National Geographic Website.2008Effects of Global Warming. Retrieved October 11, 2008, fromhttp://environment.nationalgeographic.com/ environment/global-warming/gw-effects.html

36. J. Pinkse, M. Dommisse, 2008Environmental Land Use and the Ecological Footprint of Higher LearningBusiness Strategy and the EnvironmentRetrieved September 12, 2008, from Wiley InterScience database.

37. Princeton Review.2008Green Rating Honor Roll. Retrieved September 10, 2008, from, http://www.princetonreview.com/green-honor-roll. aspx?uidbadge=07

38. K. Punch, 1998Environmental Land Use and the Ecological Footprint of Higher Learningndon: Sage Publications.

39. RecycleMania2009General Overview. Retrieved January 05, 2009, from http://www.recyclemania.org/

40. R. Spiegel, D. Meadows, 2006Green Building Materials: A Guide to Product Selection and Specification. (2nd ed.). New Jersey: John Wiley & Sons, Inc.

41. Sustainable Endowments Institute.2009Administration- Leaders-Green Report Card 2009. Retrieved March 9, 2009, fromhttp://www.greenreportcard.org/report-card-2009/categories/administration.

42. K. Turner, M. Lenzen, T. Wiedmann, J. Barret, (2006, 2006Examing the global environmental impact of regional consumption activities-Part 1: A technical note on combining input-output and ecological footprint analysis. Ecological EconomicsRetrieved October 10, 2008, from ScienceDirect database.

43. M. Wackernagel, C. Monfreda, N. Schulz, K. Erb, H. Haberl, F. Krausmann, 2003Environmental Land Use and the Ecological Footprint of Higher Learnings. Land Use PolicyRetrieved October 10, 2008, from ScienceDirect database.

44. M. Wackernagel, W. Rees, 1996Our Ecological Footprint: Reducing Human Impact on the Earth.Gabriola Island: New Society Publishers.

45. 2007, JanuaryS. Wallace, (200, January, the. Farming, Last. Amazon, the. of, Amazon, National Geographic. Retrieved October 11, 2008, fromhttp://environment.nationalgeographic.com/environment/habitats/last-of-amazon.html

46. M. Wymer, 2008UA Experts Offer 'Going Green' Advice. Retrieved February 16, 2009, fromhttp://uanews.edu/anews2008/apr08/earthday040808.htm

Chapter 9

THE ROLE OF TRADABLE PLANNING PERMITS IN ENVIRONMENTAL LAND USE PLANNING: A STOCKTAKE OF THE GERMAN DISCUSSION

Dirk Loehr

Trier University of Applied Sciences, Environmental Campus Birkenfeld, , Germany

INTRODUCTION

The idea of tradable planning permits is subject to broad discussion in some developed countries such as Switzerland (for example, see [1]), but particularly in Germany (for example, see [2]).

The German federal government intends to reduce the daily land consumption to 30 ha per day in 2020 [3]. In 13 years between 1993 and 2010, land consumption in Germany was significantly higher than 100 ha per day. In the other 5 years, the undershooting of the 100 ha mark has been mostly due to lower economic growth rates or an economic slump [4]. Particularly rural areas were affected by excessive land consumption. Almost 50% of the converted land is sealed [5].

In order to achieve the 30 ha target, there is a broad consensus about the necessity to support planning by means of economic instruments. In this discussion, tradable planning permits turned out to be the instrument of choice, at least among the scientists. In Germany, a lot of research has been underway on this issue for years now (for example, see [6]). Among others, a pilot project is also in preparation [7,8], as it was planned in the coalition agreement of the current federal government [9].

The idea of tradable planning permits stems from the concept of tradable CO_2 rights, more accurately the cap and trade system. Within the cap and trade regime, pollution rights should be limited in quantity and made tradable. Due to the cap on the pollution possible, the system is considered ecologically effective. If the mechanism is applied to the field of land use planning, the communal development plans are only legally valid if they are backed by planning permits, which have to be held by the communes. The communes

– as the planning authorities – and not the land owners are the holders of the planning permits. This is an important difference from "tradable development rights", where private-sector actors are the sellers and buyers (for example, see [10]).

Due to the trade, the scheme is also regarded as being efficient because only those actors with the lowest marginal abatement costs reduce the emissions. The permits can be bought and sold by the communes on an organized trading platform.

The cap on the permits helps to circumvent rationality traps (game theory) which otherwise would appear. In Germany, for instance, communes competed against each other to attract new inhabitants and industries in order to get more tax revenues and higher shares out of the financial equalization scheme. This competition was a race to the bottom in many cases. Among others, the results in many cases have been almost empty residential or commercial areas and high infrastructure costs. However, if a community waives the preparation of new building areas, the neighbouring municipality takes the chance.

Within the cap and trade scheme, such rationality traps might be broken up [11]: Due to the costs of the permits, only such communes whose benefits of land development exceed the costs of the permits will buy planning permits and carry out land development. If land conversion can be avoided at costs below the costs of the planning permits, communes waive the right to further development. Maybe they can also reconvert the land into a natural state. Hence, if there is no need for holding permits, such communes will sell them to other communes (for example, see [12]). If they do not sell the "free" rights, they will suffer opportunity costs. This means that communes with high marginal abatement costs (tax revenues, jobs etc.) are the buyers of the rights, while communes with low marginal abatement costs are the sellers. In the end, all marginal abatement costs equalize at the price of the tradable permits. In the trading planning permits scheme, the secondary market is the institutional heart of the mechanism.

Although it sounds quite appealing at first glance, we want to show that the application of the cap and trade scheme to land use planning is anything but self-evident and not a promising approach per se.

HYPOTHESIS: NO MAGIC BULLET

Tradable planning permits are considered to be a sort of magic bullet. On the one hand, the cap on development permits makes the system effective. On the other hand, only those communes with the highest benefits (additional taxes and shares from financial equalization schemes) carry out the development.

Communes with low opportunity costs waive the right to development. Hence the scheme is also efficient, because the planning rights are used at the locations with the highest benefits.

However, contrary to what intuition would suggest, we want to show that effectiveness and efficiency don't harmonize if the concept is applied to land use planning. In contrast, the cap and trade approach cannot meet the goals of efficiency and effectiveness at the same time ("incompatibility thesis") [13]. The argument is based on the following two statements:

- In order to be efficient, a cap and trade system needs wide system boundaries. At least in small or medium-sized countries, such wide system boundaries go hand in hand with a unified planning permit and a unit price.

- In contrast to CO_2, effective land use planning doesn't require control of a scale, but of a structure. A land use structure cannot be controlled effectively by a single planning permission with a single price.

THEORETICAL ISSUES AND REVIEW OF LITERATURE

There is a central difference between the cap and trade on CO_2 and the cap and trade on planning permits. Considering the consequences for global warming, it does not matter where the CO_2 is emitted due to the diffusion characteristics of the greenhouse gas. Hence the task is to control a scale (maximal CO_2 emissions anywhere) by capping the quantity of emissions. However, regarding land use planning, not only the scale but also the structure of land use has to be controlled. The quantity of land as a whole can hardly be extended. Instead, the relevant issue relates to changes in the structure of land use, which is for instance forestry, agriculture, industry, settlements etc. It is of central importance where the land use takes place and for which purpose. Hence CO_2 permits are a homogenous good, but land use rights shouldn't be.

Consensus: Primacy of Planning

At present, the structure of land use is controlled by the planning system. According to the proponents of the tradable development rights idea, land use planning should not be substituted but supported by the economic tool ("primacy of planning") [14].

Planning is necessary to break up a possible Nash equilibrium [11] caused by the behaviour of land owners: If, in the absence of any planning, only the willingness to pay decides about land use patterns, a spatial disaster may result and people may run into a rationality trap. If, for example, German people were allowed to realize the favoured model of the detached one-family house in green surroundings, urban sprawl would happen, with negative ecological, economic and social impacts.

At the same time, planning is necessary to protect such forms with weak financial endowments which cause important positive external effects. If no plan provided public spaces e.g. for kindergartens and schools, such forms would have to compete with actors with a high willingness to pay (e.g. banks). Hence they could not be realized. However, without such facilities, the value of the area would often be lower than with them. Good planning should consider the variety of functions of land (e.g. ecological, spiritual). Such forms of land use that move beyond efficiency and profitability are not only important for the cohesion of the social system, but in many cases also for the resilience of the ecological system (for example, see [15]). Planning has to balance the competing demands of various stakeholders, including groups with low budgets and the protection of nature.

Thesis: Efficiency Needs Wide System Boundaries

Although the primacy of planning is wide consensus, in recent debates it has been argued that tradable planning permits may counteract land use planning [13]. The "incompatibility thesis" is based on the required design of a cap and trade regime. A major justification of the system is its efficiency. The efficiency of the cap and trade system is caused by differences in marginal abatement costs:

- Those communes with high marginal abatement costs buy planning permits on the market at the lower market price. The difference is the benefits from the cap and trade system.

- On the other hand, such communes with low marginal abatement costs reduce their harmful activities and sell the free certificates on the market. The difference between market price and marginal abatement costs is the profit from abatement.

In the end, the marginal abatement costs of all the actors equal the market price of the planning permits. The higher the differences of marginal abatement costs of the acting communes, the higher the efficiency potential of the regime will be.

However, high differences in abatement costs can be achieved by a wide design in terms of space, time, participants and the objects of trade:

- In categorical terms, diverse spatial categories (living, commerce, mixed use, traffic etc.) have to be gathered in one single planning permit ("universal" certificate). This is any land for human settlement and transport infrastructure without regard to its different components;

- Regarding the market participants, the discussion is about also including individuals or NGOs instead of only permitting communes as traders;
- In spatial terms, scientists agree that the trade boundaries have to be as wide as possible (e.g. whole of Germany, no single states);
- Considering the time dimension, banking and borrowing is also discussed in order to use differences in the marginal abatement costs over the timeline.

In Germany, in the last decade the preferred design is characterized by

- A country-wide regime (although the pilot project mentioned above will only comprise selected communes) which incorporates the administrative support of the different states of the federation [16].
- A universal certificate which comprises the whole area for settlement and traffic [16]
- Banking should be allowed (at least the transfer into the following provisional period), in order to allow long-term development strategies for the communes. In contrast, there is much scepticism with regard to providing the opportunity to use the rights before owning them (borrowing) [17].
- Regarding the market participants, an extension of participants beyond the communes has not been discussed seriously so far.

Hence the efficiency potential can only be exhausted if the target is "scale" instead of "structure" (which would make sub-markets necessary). The scale target has to fix wide system boundaries (in contrast to other tradable development rights schemes; for example, see [13]). The scale target and the wide system boundaries are mutually dependent.

There is also another reason why a working trade system would not be possible without wide boundaries: Narrow markets cause high price volatility of the permits. The higher the price volatility, the more insecure the economic success of abatement activities and the less abatement activities will take place. Thus the target is to set the condition for organized trade of the permits.

Thesis: Controlling Structure by Economic Tools Needs Tightly Segmented Sub-Markets

Having shown the necessity for wide system boundaries in a cap and trade model, the next question is whether a land use structure – not a scale – can be controlled within such a system. We want to illustrate the problem within figure 1 below. The land use plan sets the allowed land use A (e.g. industry) at the maximum of C_A. The maximal land use B (e.g. housing) is limited by C_B. The marginal abatement costs (MAC) are MAC_A for land use type A and MAC_B for land use type B. For simplification purposes, the illustration doesn't include more land use types.

If a commune waives the right to development of additional sites, it has to suffer marginal abatement costs. Such marginal abatement costs are mainly opportunity costs. If, for instance, a residential area is not realized, a German municipality gets lower shares of the income tax revenues, lower property tax revenues and lower revenues out of the fiscal equalization scheme. If an industrial site cannot be realized, the opportunity costs also comprise lost business tax revenues. Also indirect effects have to be considered, such as income multiplier effects which otherwise would have been initiated by construction activities. All these interrelations and effects are quite complex and include feedbacks within the system. Basically, the scale of opportunity costs is not quite clear. Fiscal impact analysis could provide for more cost transparency, but it is in an early stage. Some fiscal impact tools used so far for residential areas turned out to have quite different performance; for industrial areas no reliable fiscal impact tool is available so far. Also within the above-mentioned pilot model of tradable development rights the development of reliable fiscal impact analysis tools is acknowledged to be important in order to get a better idea about the marginal abatement costs.

However, in the subsequent figure we assume, contrary to the facts, that there is an accurate idea about the volume of the marginal abatement costs of land use type A and B. Hence we can derive a mathematical function of MAC, being dependent on the scale of land use of the different types.

First, let's assume that the caps for land use type A (C_A) and B (C_B) are set according to the land use plans. We assume that the land use planning also properly computes the marginal damages (MD), which are illustrated with the dotted line for both types of land use. With this theoretical "trick" we can take into account that planners care for quantity as well as for quality of land use (for example, see [17]). Therefore, the planning target (C_A and C_B) corresponds perfectly with the intersection of marginal abatement costs and marginal damage of land use type A (E_A) and B (E_B).

Moreover, theoretically the marginal damage of land use type B can be expressed in equivalents of the marginal damage of land use type A. Such equivalents are useful for the definition of a universal cap. Analogous equivalents are also used in the greenhouse gas emission permit schemes. In the Kyoto regime for instance, global warming potentials (GWPs) are used in order to express the global warming potential of the other greenhouse gases in relation to CO_2, whose GWP is standardized to 1. Hence, if caps for different land use types should reflect such equivalents (for simplification purposes a linear function is used below), we get for instance a function such as:

$$C_B = e \times C_A \tag{1}$$

The total cap results by aggregation of the caps of the individual land use types:

$$C_{A+B} = C_A \times e + C_A \tag{2}$$

It is important to note that the equivalents only reflect an average consideration. However, different communes may have different structures of land use; thus the equivalents don't represent their individual situation.

In the subsequent diagram, the added marginal abatement costs (MAC_{A+B}) show the aggregate demand curve, and the aggregated cap (C_{A+B}) shows the aggregate supply curve of all land used for settlement and traffic, set by the planning authorities. The intersection point determines the unit price of the universal development right P* [13]. The illustration holds true for an individual municipality as well as for the aggregation of municipalities.

The figure shows why a unit price causes economic incentives to violate the land use plans:

Regarding land use type A, the mayor in charge will extend the land use until the intersection point (I_A) of the unit price P* and the marginal abatement costs MAC_A. From this point on, the costs for additional planning permits exceed the benefits from additional land use. However, regarding the land use plan (and the marginal damage), more development of land use type A would be possible (up to E_A and C_A respectively). Insofar there is a loss of welfare, indicated by the gap between M_A and C_A. In order to support the land use planning, price P_A would be necessary instead of the unit price P*.

Looking at land use type B, the mayor increases land development until his/her benefits of additional development MAC_B equal the unit price of the development P* (in point I_B or M_B respectively). However, this is much more than the land use plans have fixed (C_B). If the caps reflect the intersection point E_B of marginal damage function MD_B and marginal abatement costs MAC_B, this point I_B is also far beyond efficiency (in E_B). In order to get an effective and efficient result, price P_B would have been necessary. From the figure above we may derive two important results:

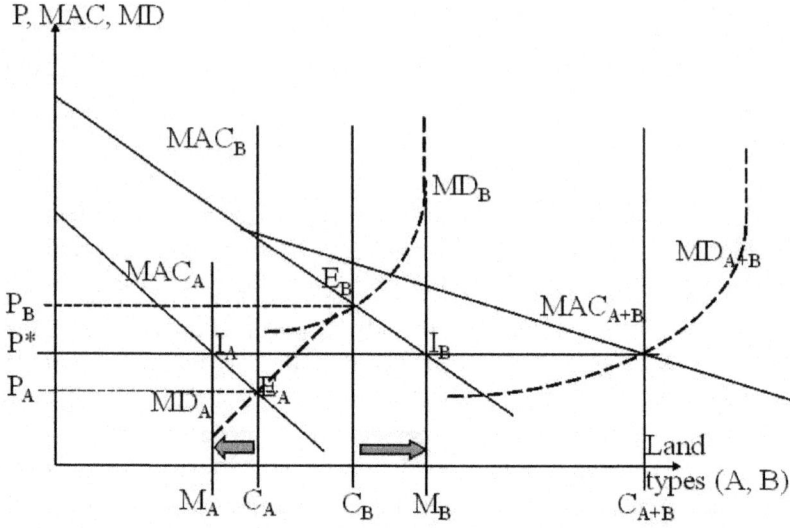

Figure 1: Aberrations with a universal planning permit (adapted from [13])

The trading model, which is based on wide system boundaries and a single tradable planning permit, is not able to support land use planning. With a unit price, it is only able to control scale but not structure.

Nonetheless, the supporters assert effectiveness. Obviously, they refer to the CO_2 blueprint and only consider the control of scale, but ignore the necessity to control structure. There is no assessment of the welfare losses which are caused by overshooting and undershooting of the planning targets so far (differences between $M_{A,B}$ and $C_{A,B}$). Therefore, also no proper statement about the net efficiency gains (efficiency gains minus the welfare losses, due to overshooting and undershooting) of the cap and trade system can be provided.

Sometimes, supporters suppose that the deviations and aberrations are negligible. Depending on the price of the permits, the position of the caps and the marginal abatement costs, the aberrations and welfare losses can be randomly quite high or low. There is no evidence for a correction mechanism, which might be able to reduce such aberrations systematically. Hence, overshootings and undershootings probably don't equalize in aggregation. The system is as effective as a poor marksman who is currently missing the target in different directions. Aggregating the shooting errors doesn't turn the poor marksman into a champion.

The "incompatibility thesis"

Against this background, we can describe the central system conflict as follows:

The rationale of the cap and trade regime is efficiency. However, efficiency can only be achieved within a wide design of the system, among others based on a universal certificate (including all categories of land for human settlement and transport infrastructure);

In contrast, the required primacy of planning can only be maintained within a tightly structured system. In contrast to the controlling of the scale of CO_2 emissions, the settlement structure cannot be controlled by one cap (certificate) and one price, but needs a diversity of prices with a diversity of certificates.

If a structure is "treated" with a single price, welfare losses have to be expected due to overshooting and undershooting of the planning targets. Hence the net efficiency gains and net welfare effects of a cap and trade regime for land use policy are not clear at all.

The supporters of the tradable planning permits claim that planning should impose a correcting action if necessary. However, our argument is that precisely due to the counteracting economic signals, planning cannot impose such an action.

The problem could theoretically be solved by a variety of tightly segmented sub-markets with diverse price settings. Sub-markets which adapt the categories of land use planning (market segmentation in categorical terms) have been discussed e.g. by Henger and Schröter-Schlaack [17]. Spatial boundaries which limit the trading rights to regions with similar protection status (market segmentation in spatial terms) have been addressed e.g. by Williams [18], Walz et al. [16], Henger and Bizer [19]. However, in most countries such markets would be too small and thus inefficient [17]. At least in the German discussion, the conflict of goals which appears in the cap and trade scheme was decided in favour of efficiency and against market segmentation. Within a system with wide boundaries, such an incompatibility could perhaps be avoided in a few countries with high population and centralized land use planning systems, such as China. However, so far there is no sound research about the minimum size of such markets and the requirements for the land use planning system.

Addendum: Initial Distribution

Within a cap and trade system, the initial distribution might be done by auction or by providing the permits to the communes without costs (according to alternative allotment formulas). Simulation experiments with German municipalities related to the cap and trade scheme also showed that the initial distribution of rights is quite a critical issue, which may endanger acceptance [20]. In order to guarantee the acquis of the communes, discussions have so far favoured "grandfathering" schemes, in which the status quo of land consumption is not touched. However, such grandfathering schemes are probably less efficient than auction schemes [13, 17].

RESULTS AND DISCUSSION

Basically, the required primacy of planning is not compatible with the system of tradable planning permits. If the basic idea of capping planning permits should be kept, a redesign of the regime is necessary.

Cap and Auction

In order to support the planning system ("primacy of planning"), the planning permits should best be defined in a tight manner which is in line with the categories of the planning system. For instance, if planners are thinking in categories such as residential, industrial or mixed areas, traffic etc., the planning permits should also follow these categories. This also facilitates the handling of the system. Planning permits should basically be mandatory for all sorts of developed areas. For instance, recreational space may have also negative ecological impacts.

The caps could be administered on different administrative levels, e.g. at the level of the states, even at the level of regions (or in other countries: at county level). However, the administration needs a certain human capacity. In Germany, the 30 ha cap might be broken down into lower administrative levels without problems.

However, within a tight definition of a variety of planning permits at a low administration level (e.g. region), the "markets" for each right would be quite narrow. Hence the allocation mechanism shouldn't be based on the secondary market (trade – "horizontal coordination") but on the primary market, namely auctions ("vertical coordination"). The auction of planning permits to the communes could be done periodically. Giving the focus to auction doesn't mean a complete ban on trade but a reduction of its significance. Due to the tight design of the sub-markets, organized trade wouldn't be possible anyway. Instead, over-the-counter trade would be feasible, regulated by the planning permits administration. Within the proposed design, different types of planning rights (residential, industrial etc.) were auctioned and traded at different prices.

The administration should also guarantee that it will take back the planning permits for a fixed price (based on the auction price). Thus communes could also think about changing the land use plans and the redevelopment of shrinking areas into the natural state. Deconstruction would be encouraged, because the communes could be sure about the compensation.

A system which is based on auction might be designed tightly with regard to space, time, participants and objects of the design. The system may work with only a few participants. It may work even at regional level. Within the auction, the development rights are allocated to those communes that can make the best use of them. Moreover, without the overshooting and undershooting of the cap and trade regime, welfare losses also are avoided.

Despite the segregation of sub-markets, the system is also efficient. However, in terms of efficiency it is not clear if such a cap and auction system will or will not compete with a cap and trade system, which is based on a single, universal development right. Nonetheless, if in doubt, recognizing the primacy of planning within the conflicts of goals means subordination of efficiency. From a system theory point of view, economic efficiency shouldn't be a guiding value [15] of superior significance anyway, at least not in land use management. Instead, the guiding value of efficiency has to be balanced with other guiding values.

Completion By Means Of a Financial Equalization Scheme

The regime sketched out above has to face some serious counterarguments relating to political viability: In an auction, the powerful communes with a high willingness to pay will prevail. Moreover, the financial situation of the communes would be even more strained and the municipalities' acquis would be encroached (see section 3.5.). Hence the question is how to increase the acceptance of a cap and auction model.

On the one hand, certain transition regulations such as free development rights for existing settlements would certainly be helpful, but not nearly enough on their own.

On the other hand, the view of the discussion about climate policy might be promising. Within the Kyoto regime, it has not been possible to put in place effective caps, mainly because the problems of distribution and "climate justice" have not been solved yet [21]. The regime was based on "grandfathering". Hence those countries with the most aggressive occupancy of the atmosphere and most responsibility for the climate problem got most rights. This was considered as being unjust by countries with developing and emerging economies. Basically, the same holds true in terms of land use permits.

However, many of the objections mentioned above could be countered by establishing a redistribution mechanism. Within such a redistribution mechanism, the money paid by the communes in the auction could be firstly collected in a fund ("land trust"), which is administered by the affected communes. Second, the money is redistributed to the communes, preferably according to the number of their inhabitants (other redistribution keys are also possible). In this respect, all inhabitants are considered as "co-owners" of the planning permits, and thus they should participate in equal shares from the revenues of the auction. Considering the CO_2 emission trading schemes, this idea has been popularized by Peter Barnes [22]. Applied to land use planning, a similar redistribution scheme has already been suggested by Krumm [23]. However, his proposal was based on a price-steering basis: Basically, communes should be charged for any new land conversion using a fixed rate per square meter. The money should be pooled in a fund and redistributed to the communes, preferably according to the number of citizens.

If redistribution to the communes were carried out according to the population, the payments into the "land trust" would be according to the land used per capita, whereas the redistribution would be according to the average use per capita. Hence, besides the cap, an additional incentive for a sustainable land use is implemented:

If the actual land use per capita is higher than the average land use, the commune in charge is a net contributor to the "land trust";

If the actual land use per capita is lower than the average, the responsible commune is a net beneficiary;

If the actual land use corresponds to average land use, there is no difference compared with the status quo.

Because every commune tries to get net benefits out of the land trust, there will be a current dynamic incentive to carry out efficient and effective land use management. In terms of microeconomics, the dynamic incentive is pushed by the substitution effect, whereas the income effect is eliminated by the redistribution scheme ("Slutsky equation", see [24]).

Moreover, within this redistribution mechanism, an average access to the planning permits is granted, also for financially weak communes. The redistribution mechanism serves as an ecological financial equalization scheme between the municipalities. Not unlike a lease mechanism, communes with land consumption rates above average pay to communes with land consumption rates below average.

The effects of the redistribution system are far reaching. To mention just some of them:

Currently, for instance, some German communes can take some "fiscal rents" due to their location, at the expense of other municipalities. This holds true particularly for the communes in the wealthy commuter belt of bigger cities ("Speckgürtel"). They benefit from the migration out of the bigger cities (e.g. young families), which are "bleeding". In such peripheral communes, land prices are often lower and the environmental conditions are often better than in the big cities. However, a great deal of the attractiveness is caused by uncompensated spillovers. According to the central locations principle [25] the bigger cities provide a variety of public goods at the commuter belt's benefit. Thus urban sprawl is fuelled, and the financial performance of bigger cities gets weaker and weaker. However, in the proposed regime, the whole fiscal surplus will be skimmed off dependent on the type of auction. The willingness to pay of the commuter belt's communes includes the expected fiscal rents (from spillovers). The fiscal rents are redistributed to all communes according to the number of the people, also to the bigger cities. Due to the higher density of population, the redistribution scheme will compensate the bigger cities for their efforts.

The model is applicable in situations of growth as well as in shrinking areas. In aging societies such as Germany, in particular rural regions are affected by shrinking. However, land conversion and land consumption is highest ex urbia. Urban sprawl turns out to be luxury which is increasingly difficult to finance. The redistribution model may stimulate migration to more compact settlements, with a higher supply of public goods. The system would provide an incentive for renaturation measures in rural communes. This would have positive side effects, considering e.g. vacancy rates and the value of existing properties.

By skimming off the rents from certain types of land use, communes get more indifferent towards land use alternatives. On a regional level, coordination between municipalities and the allotment of certain functions (industry, tourism etc.) towards different communes is easier than today. Thus, integrated approaches of regional development might be put in place without high resistance of the communes affected.

With regard to the technical implementation of the system, some minor problems have to be solved. For instance, in order to create equal conditions in the auction, the communes should pay into the land trust in the same "logical second" as the redistribution happens. This means the communes are only charged or rewarded by the balances (net position of pay-in and pay-out). Moreover, it has to be figured out on which administrative level the system should be applied. Basically, the redistribution mechanism should be tied to the scope of the cap and auction scheme.

One should be clear about the fact that no money for natural protection would be raised within the redistribution scheme. However, modifications are possible: If, for instance, a natural park as a common public good has to be financed, the redistribution could be carried out after first deducting the expenses for covering the park. Such decisions depend on the land trust and the planning authorities. A legal basis for the cooperation arrangement is necessary.

The proposed model may be appealing, but it is not a "silver bullet". The framework has to be completed by other instruments. For instance, the price of real estate may rise due to successful capping of planning rights. Thus, access problems for socially weak groups might be caused. Hence a suitable land taxation system which transfers shares of land rents and land value to the community would be desirable for example.

CONCLUSIONS

The concept of tradable planning permits transforms the idea of the CO_2 cap and trade regime to spatial planning. Analyzing the tool, we have at least to refer to effectiveness (planning, ecology), allocation (economy) and distributional aspects (social).

Regarding effectiveness, there is a broad consensus about the primacy of planning. Any economic tool should support planning instead of substituting it. However, planning land is not the same as planning the maximum permissible load of CO_2 in the atmosphere. The former requires a planning of structure, the latter a planning of scale, since it is irrelevant where the emission takes place.

Planning the structure of land use cannot be supported by a unit price, as a result of a universal certificate for all types of land use (for settlement and traffic). In contrast, a variety of sub-markets are necessary, with a different price setting. Meanwhile, more and more planners are also becoming sceptical about the supporting effects of a cap and trade regime.

Supporting the planning of a structure within a variety of sub-markets may be inferior compared with the efficiency of a cap and trade system with wide system boundaries. On the other hand, efficiency losses due to overshooting or undershooting might be avoided. The efficiency losses might be minimized by auctioning the permits to the needy communes on the primary market ("vertical allocation"). Although trade shouldn't be forbidden, an organized secondary market is dispensable (subordination of a "horizontal allocation mechanism"). Moreover, both systems would have to prevent strategic acquisitions of permits (impediment of development in other communes by an artificial shortage of supply), e.g. by a current devaluation of the permits.

Regarding the blueprint of CO_2 trade, a comprehensive arrangement on a global scale has so far failed due to distribution disputes. Also a cap and auction system for planning permits wouldn't be acceptable particularly for communes with a weak financial endowment if there were no correction. This is the reason why the cap and auction regime should be completed by a redistribution mechanism which is based on equal stakes in the scarce land use opportunities.

However, more research is necessary in order to deal with the details of the counter-proposal outlined in this article. So far, in Germany politics has supported the cap and trade approach; as has the allocation of research funds. Critics who pointed out the incompatibility between effectiveness and efficiency in the cap and trade approach have been pushed aside. This also holds true for the combination of caps, auction and redistribution, which couldn't be assessed so far. However, in experimental simulations the acceptance of the cap and trade regime among the practitioners was obviously not very high. Among others, the results of the cap and trade game turned out to be quite sensitive in terms of an increase of the complexity of the framework [20]. In contrast, at least the redistribution approach of Krumm was highly accepted (here, basically, also no fiscal impact assessment is necessary) [26]. Maybe it is time to widen the scope of the research paradigm to extend beyond the cap and trade regime.

REFERENCES

1. F. Zollinger, I. Seidl, 2005Flächenzertifikate für eine nachhaltige Raumentwicklung?- Ein Konzept für Baden-Württemberg und Erkenntnisse aus der Übertragung auf die Schweiz. Informationen zur Raumentwicklung 4/5273280

2. U. Kriese, 2005Handelbare Flächenfestsetzungskontingente-Anforderungen an ein Mittel zur Beendigung des Landschaftsverbrauchs. Informationen zur Raumentwicklung 4/5297306

3. Federal Ministry for the Environment, Natural Conservation and Nuclear Safety1998Nachhaltige Entwicklung in Deutschland- Entwurf eines umweltpolitischen Schwerpunktprogramms, Berlin, 147 p.

4. G. Penn-Bressel, 2011Flächenneuinanspruchnahme- Wirkungen auf Umwelt, Städtebau und Ökonomie. Wirtschaftsdienst 11800802

5. S. Frerichs, M. Lieber, T. Preuß, 2010Flächen- und Standortinformationen erheben und bewerten- Methoden und Konzepte für ein nachhaltiges Flächenmanagement. In: Frerichs, S, Lieber, M, Preuß, T, editors. Flächen- und Standortbewertung für ein nachhaltiges Flächenmanagement. Berlin: Difu 1127

6. Homepage of the "Refina" research programme, funded by the federal government: http://www.refina-info.de.Accessed: 2012Mar 15.

7. Federal Ministry for the Environment, Natural Conservation and Nuclear Safety2010Environment research plan (Umweltforschungsplan) 2010. Project Z 691Project 3710

8. Homepage of the project "Forum Flächenzertifikate".Available: http://www.ufz.de/index.php?de=21103Accessed 2012Mar 15

9. C. D. U. C. S. U. F. D. P. , 2009Wachstum, Bildung, Zusammenhalt, Coalition Agreement. Berlin.

10. LeJava, J P2009Transfer of Development Rights in New Jersey- A background paper, New Jersey. Available:http://www.dvrpc.org/TDR/pdf/200910_LeJava_Background_Paper.pdfAccessed 2012 Mar 15.

11. Nash J F1950Non-cooperative Games; Ph.D. Thesis; Princeton University: Princeton, USA. Available:http://www.princeton.edu/mudd/news/faq/topics/Non-Cooperative_Games_Nash.pdf.Accessed   2012 Mar 15.

12. R. Krumm, 2004Nachhaltigkeitskonforme Flächennutzungspolitik-Ökonomische Steuerungsinstrumente und deren gesellschaftliche Akzeptanz, IAW research report, Tübingen: IAW. 136 p.

13. D. Loehr, 2006Handelbare Flächenausweisungskontingente: Eine gute Idee auf Abwegen, in: Zeitschrift für Umweltpolitik und Umweltrecht 4529544

14. J. Bovet, 2006Handelbare Flächenausweisungsrechte aus Steuerungsinstrument zur Reduzierung der Flächeninanspruchnahme. Natur und Recht 6473479

15. H. Bossel, 1998Globale Wende- Wege zu einem gesellschaftlichen und ökologischen Strukturwandel, Munich: Droemer. 464 p.

16. R. Walz, et al.2005Gestaltung eines Modells handelbarer Flächenausweisungskontingente unter Berücksichtigung ökologischer, ökonomischer, rechtlicher und sozialer Aspekte, Final Report. Research Project, funded by the Federal Environmental Office, Project 203Dessau-Roßlau. 170 p. Available: http://opus.kobv.de/zlb/volltexte/2009/7870/pdf/3839.pdf.Accessed 2012 Mar 15.

17. R. Henger, C. Schröter-Schlaack, 2008Designoptionen für den Handel mit Flächenausweisungsrechten in Deutschland, Land Use Economics and Planning- Discussion Paper, 08-020802University of Göttingen, September. Available: http://www.uni-goettingen.de/en/80714.htmlAccessed 2012 Mar 15.

18. Williams R C2003Cost-Effectiveness vs. Hot Spots: Determining the optimal size of emissions permit trading zones. University of Texas at Austin Working Paper.

19. R. Henger, K. Bizer, 2008Tradable Planning Permits for Land-use Control in Germany. Land Use Economics and Planning- Discussion Paper 08-010801

20. C. Küpfer, et al.2010Handelbare Flächenausweisungszertifikate-Experiment Spiel.Raum: Ergebnisse einer Simulation in 14 Kommunen. Naturschutz und Landschaftsplanung 4223947

21. L. Wicke, 2006Das Versagen des Kyoto-Protokolls in seiner jetzigen Form und seine strukturelle Weiterentwicklung. Zeitschrift für Sozialökonomie 15039

22. P. Barnes, R. Pomerance, 2000Pie in the Sky- The Battle for Atmospheric Scarcity Rent, Washington. Available: http://community-wealth. org/_pdfs/articles-publications/commons/paper-barnes-pomerance. pdfAccessed 2012 Mar 20.

23. R. Krumm, 2002Die Baulandausweisungsumlage als ökonomisches Steuerungsinstrument einer nachhaltigen Flächenpolitik. IAW Discussion Papers 7. Available: http://iaw.edu/iaw/De:Publikationen:IAW-Reihen:IAW-Diskussionspapiere:2002Accessed 2012 Mar 20.

24. H. Varian, 1992Microeconomic Analysis, 3rd ed., New York: W.W. Norton & Company Inc. 506 p.

25. W. Christaller, 1933Die zentralen Orte in Süddeutschland. 2nd ed. 1968. Darmstadt: Wissenschaftliche Buchgesellschaft. 331 p.

26. T. Preuß, et al.2007Perspektive Flächenkreislaufwirtschaft, 3Neue Instrumente für neue Ziele. Berlin / Bonn: Difu. 109 p.

Chapter 10

SPATIAL ANALYSIS FOR FLOOD CONTROL BY USING ENVIRONMENTAL MODELING

Alireza Gharagozlou, Hassan Nazari, Mohammadjavad Seddighi

Geomatics College of National Cartographic Center of Iran (NCC), Tehran, Iran

ABSTRACT

To create the final spatial information and analysis, flood hazard maps and land development priority maps and information, data for the flood events to 2009 in north of Iran were incorporated with using Geo-spatial Information System data of physiographic divisions, geologic divisions, land cover classification, elevation, drainage network, administrative districts and population density and environmental parameters modeling. Special analysis also attention was paid to population density for the construction of the land development priority map and using satellite image analysis to determine land use changes and analysis of geo-spatial information, because highly dense populated areas represent the highly important urban and industrial areas. While geo-information technology offers an opportunity to support flood management adequate geo-spatial information is a prerequisite for sustainable development, but many parts of the world lack adequate information on environmental resources. Such information providing, which serves as an important tool for decision-making in land use planning, can help provide effective information to natural disaster management. This paper develops a framework for flood control and begins with some general comments on the importance of land use planning and outlines some current environmental issues and then presenting environmental models to use in disaster management plan by using GIS and remote sensing results. Flood control is a complex problem that requires cooperation of many scientists in different fields. The article also discusses the role that geo-information and environmental planning and GIS and remote sensing technology play in disaster management control to reduce negative impacts of flood and present proper alternatives for developing of Gorganrood in the north of Iran. Advanced high-resolution sensor technology has provided immense scope to the decision makers for analysis of flood and damages details using GIS and remote sensing.

INTRODUCTION

This article begins with some general comments on the importance of land use planning and outlines some current environmental issues. It also highlights the connection between land use planning and sustainable development and the discussion describes several key methods of resource identification, with particular emphasis on existing potential of geo-information technology that offers an opportunity to support disaster management: floods and environmental impacts assessment and natural disaster in national level. The article also discusses the role of geo-information data in promoting geographic information system use. By attention to natural disasters in Iran especially flood in Golestan in the North of Iran proper assessment of flood by using environmental development models and GIS and with attention to Sustainable development approach and disaster management are presented. The article offers proposed models that illustrate how GIS and remote sensing data can be used in land use planning programs that take a sustainable development approach [1] and disaster management (flood). Excessive land use and increased human impacts have imposed significant pressures on the environment worldwide. These effects are increasingly noticeable from a scientific and technical viewpoint. Future development should proceed on the basis of proper land use planning, with minimum destruction of the environment because impacts of human activities results natural disasters in some area [2]. Planning assessments must therefore consider environmental issues and natural disaster (flood) and use environmental and dereferenced information to refine decisions. Gathering information reveals the available potential of the environment; development planning at the nationwide level can help decision-makers identify resources and target their future scientific studies to reach sustainable development. Moreover, each minute, 5.6 hectares of forest are being destroyed and some other human activities cause disaster such as flood in some parts of the world.

MATERIALS AND METHODS

Floods are one of the most common hazards in the world also in the some parts of north of Iran. Flood effects can be local, impacting a neighborhood or community, or very large, affecting entire river basins and multiple states [3]. However, all floods are not alike. Floods themselves average four billion dollars annually in property damage alone. Some floods develop slowly, sometimes over a period of days. But flash floods can develop quickly, sometimes in just a few minutes and without any visible signs of rain. Flash floods often have a dangerous wall of roaring water that carries rocks, mud, and other debris and can sweep away most things in its path. Overland flooding occurs outside a defined river or stream, such as when a levee is breached, but still can be destructive. Some general reasons of flood include: weather related reasons: heavy rainfall, duration of precipitation, sudden snow melting and physical conditions: soil variety, slop of lands, land degradation and human activities: deforestation, misusing of land and transforming to grasslands or agricultural area, misconstruction of roads, bridges, dams and environmental situations. Flooding can also occur when a dam breaks, producing effects similar to flash floods. Be aware of flood hazards no matter where you live, but especially if you live in a low-lying area, near water or downstream from a dam. Even very small streams, gullies, creeks, culverts, dry streambeds, or low-lying ground that appear harmless in dry weather can flood every state is at risk from this hazard [4]. Some scientists think the major problem about natural disaster and flood is in the improper exploitation of land [5]. By using process of plan compilation with a land use planning approach some important negative impacts that cause flood is under our control.

Identification of Land Resources for Planning

Statistics and sampling, conversion of the aerial photos, satellite images and topographic maps, automatic conversion of aerial photos and satellite images and data of remote sensing, geographic information systems are different methods of identification of resources. One of the objectives of this study is to utilize GIS data to construct a set of GIS data, a flood hazard map, and land development priority map to help the responsible authorities develop, design and operate flood control infrastructure and prepared aid and relief operations for high-risk areas during future floods. In recent years the combination of 3D-laser scanning and side-scan can be very beneficial for mapping complicated water side areas; the two systems are complementary [6]. To geo-reference the relative location, GPS positioning required. It should be clear that presenting an environmental development model to be used in a GIS for natural disaster management has a lot of restrictions and limitations [7] whose description would lead too far here. Some factors that have been considered in presenting the model include; industrial sites, transportation networks, weather and climate data, landform, elevation, slop, geology, bedrock, soil, water resources, vegetation, installations and buildings, energy transmission stations, natural resources, gardens, forests, parks, etc. the priority of the mentioned parameters are different in the model[8]. It is clear that north of Iran and Golestan has an environmental development context and is under the interactive effects of the large region. Also it is thus impossible to correctly analyze the environmental conditions for natural disaster management without considering the social and economic activities in this district.

Modeling

Environmental modeling is a complex problem that requires cooperation of many scientists in different areas. In this paper, the architecture and results of environmental modeling and using satellite image processing and GIS for Flood control is presented. Set out below are mathematical linear models for flood management in Golestan. Flood inundation modeling requires distributed model predictions to inform major decisions relating to planning [9]. Present flood model integrates GIS with the environmental modeling and greatest daily of precipitation from1995 to 2009 to determine improper area for development. A linear mathematical model for flood management has been Introduced because proper planning based on environmental potentials cause reduce risk potentials of flood in future and this linear model with attention to planning results sufficient for development impacts on disaster. FF refer to specific model with environmental planning approach for flood management and attention proper land use planning in the region that present location of improper area for development. Predicting the river's flood is one of the important factors for design of dams and hydraulic structures and regional and urban development planning. As geo-information data also used in flood management, many problems occur in flood estimation. One of the methods for planning is determination improper area for development by environmental modeling, statistics and using GIS/RS technology [10].

$$R = S\,(4,5,6,7,8) + AS\,(1,2,3,7,8) + H\,(1,2,3,4) + B\,((x,y) > RR)$$
$$RR = R1\,(V1+M1) + R2\,(V2+M2) + R3\,(V3+M3) + ... \ FF = S(5,6) + H(5,6) + QA(2,3,4) + MA(1,2,3)$$
$$+ WS(5,6) + SO(1,2,4,5,6) + SW(1,2,3) + NI(1,2,3) + HP(1,2,3) + HBU(1,2,3) + HBR(1,2)$$
$$+ So\,(3,4,5,7,8,9,10,11) + Sd\,(4,5) + Prc\,(1,2)$$

where: S is slope, H is height or altitude, A is aspect, QA is fault line, MA is distance from ravine areas, WS is wind speed, SO is soil components, SW is distance from subterranean water resources, NI is distance from industrial sites and HP is historical landmark, HBU is Distance from urban habitat and HBR is distance from rural habitat and Lo, Ma(R-year) is the maximum precipitation based on geographical location. For purposes of the linear models, the terms used have the following definitions: "Slope" (S) includes six classes: 0 to 2% (class 1), 2 to 5% (class 2), 5 to 8% (class 3), 8 to 12% (class 4), 12 to 15% (class 5), and more than 15% (class 6). "Height" (H) includes six altitude classes: less than 1000 meters (class 1), 1000 - 1200 meters (class 2), 1200 - 1400 meters (class 3), 1400 - 1600 meters (class 4), 1600 - 1800 meters (class

5), and more than 1800 meters (class 6). "Distance from ravine areas" (MA) includes four classes: less than 50 meters (class 1), 50 - 300 meters (class 2), 300 - 500 meters (class 3), and more than 500 meters (class 4). "Subterranean water resources" (SW) divides resources into four classes, based on distance to the water resource: less than 100 meters (class 1), 100 - 500 meters (class 2), 500 - 1000 meters (class 3), and more than one kilometer (class 4). "Distance from industrial sites" (NI) includes three classes: less than 5 kilometers (class 1), 5 - 10 kilometers (class 2), and 10 - 20 kilometers (class 3). "Distance from Urban Habitat" (HBU) divides 4 four classes: less than 5 kilometers (class 1), between 5 to 10 kilometers (class 2), between 10 - 20 kilometers (class 3) and more than 20 kilometers (class 4), "Historical landmark" (HP) divides historical places into four classes, based on how far away they are located: less than 5 kilometers (class 1), 5 - 10 kilometers (class 2), 10 - 20 kilometers (class 3), and more than 20 kilometers (class 4), Distance from rural habitat "HBR" divides rural area and around this to 4 classes: less than 2 kilometers (class 1), 2 - 4 kilometers (class 2), 4 - 8 kilometers (class 3) and more than 8 kilometers class 4 and Prc is precipitation in mm in 7 classes more than 2000, 1800 - 2000, 1200 - 1800, 800 - 1200, 500 - 800, 200 - 500, 50 - 200 and less than 200 mm and So is soil construction in 11 classes and Sd is soil depth in 5 classes include more than 180, 120 - 180, 60 - 120, 30 - 60, less than 30 cm. With attention to linear model of flood control and data analysis and using GIS the result of analysis are as maps presented that are presented in Figures 1-4. Satellite data also can be effectively used for mapping and monitoring the flood inundated areas, flood damage assessment, flood hazard zoning and post-flood survey of rivers configuration and protection works. Analyzing the satellite images reveal a noticeable reduction of forestlands in north of Iran due to the expansion of the urban limits misuse from these area. The other fixed natural resources of the region too have been overused resulting in environmental destruction of the area. The amount of residential areas during 1995 and 2009 show an 8% growth while there is no increase in the number of forestlands. There are several definitions of sustainability in the urban forestry sector that we attention in this paper based on analysis results and conditions in urban area. The amount of forestlands declined about 1 hectare every year and open areas have been reduced thus leading to the conclusion that most of the construction activity took place in forestlands and results suiting with other experiences in this filed.

CONCLUSIONS

In this paper Innovations in the filed of environmental modeling which are based on natural disaster management for flood in Iran and using proper models for analysis in GIS are presented. Geo-information technology offers an opportunity to support disaster management and floods as the natural disaster management. Gorganrood rivers waterways trails as a pedestrian pass with a stream of spring and rain which also can hold heavy rain water as a flood control waterway. One of the obvious and prominent aspectsof innovations in this paper, are the models that can integrate between Geo-information technology and environmental modeling and natural disaster management. Analysis positioning of improper locations by using GIS/RS technology for development area determination based on the environmental capacities with a flood control approach. At the same time, by using GIS necessary analysis to find flood risk in the region and impacts of flood on natural and human facilities are presented. Choosing proper linear models based on environmental capacity with a flood management emphasize determining the ecological potentials of the area and using GIS is the important point of this paper. The joint application of GIS and environmental modeling and using remote sensing technology can help land use planners apply optimal development planning guidelines. The other key idea we suggest here is the need to compare the results of these analyses with future development plans. Comparing the natural potential of the territory with predicted development plans can result in better decision making to reduce the cost of flood in rural and urban area. The use of GIS technique during the last decade are increasing being applied for identification of natural resources but the practice of analyzing the development models with the use of GIS/RS in development planning for flood management is a new experience.

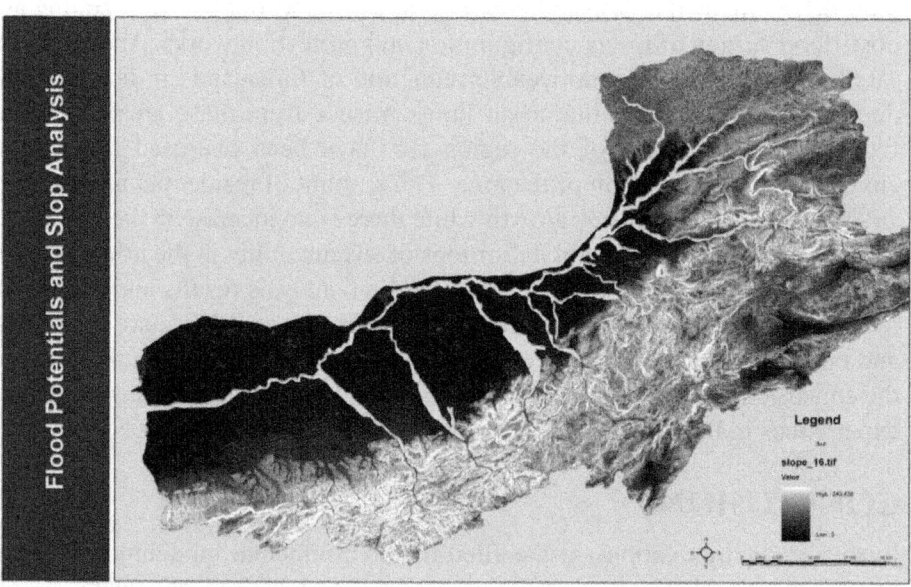

Flood Potentials and Slop Analysis

Figure 1: Spatial Analysis for determination flood risk area and improper area for development using flood modeling in north of Iran

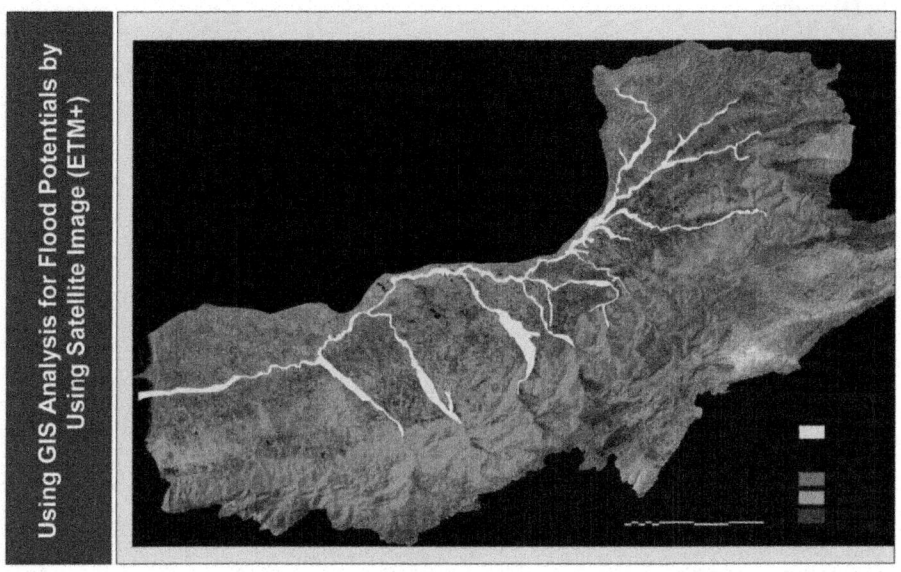

Figure 2: Satellite image processing (ETM+) in Golestan, Gorganrood, and risk assessment and modeling for flood control.

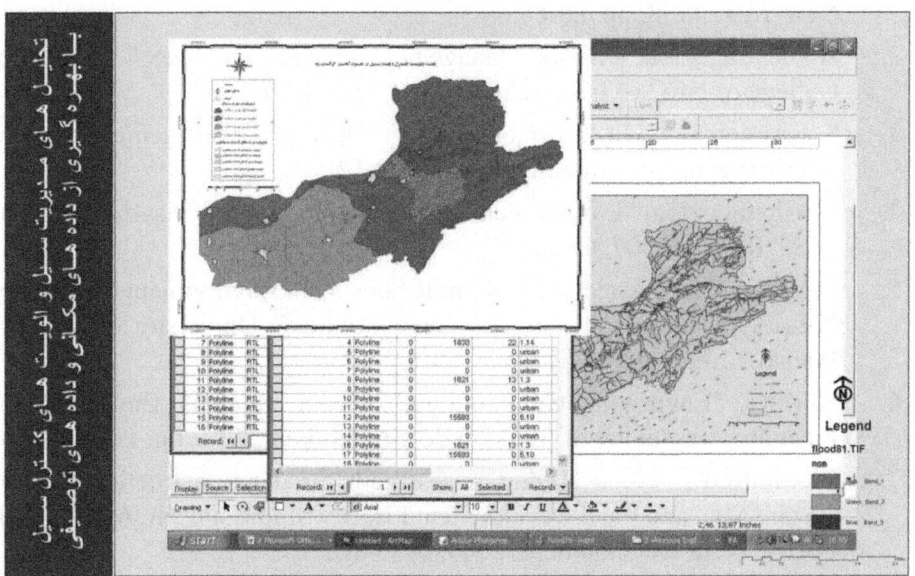

Figure 3: Districts risk classification for flood control by using GIS.

Figure 4: Land reform and erosion and cities under flood risk.

REFERENCES

1. J. Hardisty, D. M. Taylora and S. E. Metcalf, "Computerized Environmental Modeling: A Practical Introduction Using Excel," Wiley, New York, 1993, pp. 80-88.

2. A. Gharagozlou, "Crisis Management (Flood) and GIS," Geomatics College of NCC of Iran, Tehran, 2010, pp. 23-29.

3. S. Andera, "Geo-Information for Disaster Management," GIM International, Amsterdam, 2005, pp. 10-12.

4. A. Gharagozlou, "Environmental Planning for Natural Disaster by Using GIS," ISPRS, 2007, China.

5. E. Wolk and B. Zagajewski, "Remote Sensing of Environment Laboratory, Faculty of Geography and Regional Studies," University of Warsaw, Warsaw, 1999, pp. 44-57.

6. Fairley; "Environmental Planning," Department of Environmental Planning, University of Stratchlyde, Stratchlyde, 2001, pp. 111

7. S. kimitrero, "Flood Hazard Assessment for the Constructionof Flood Hazard Map and Land Development Priority Map Using NOAA/AVHRR DATA," GIS Development, 2006, pp. 3-12.

8. Heywood LAN, S, Cornelius and S. Carver, "Cornelius; -An Introduction to Geographic Information Systems," chap., 1998, pp. 2-5.

9. K. Clayton, "The Land Form Space, Environmental Science for Environmental Management," Longman, London, 1990, pp. 198-222.

10. F. Ferrini, "Sustainable management techniques for trees in the urban area, Journal of Biodiversity and Ecological Sciences," IAU University, Tonekabon, 2010, pp. 1-19

Chapter 9

SOIL EROSION PREDICTION USING MORGAN-MORGAN-FINNEY MODEL IN A GIS ENVIRONMENT IN NORTHERN ETHIOPIA CATCHMENT

Gebreyesus Brhane Tesfahunegn,[1,2] Lulseged Tamene,[3] and Paul L. G. Vlek[1]

[1]College of Agriculture, Aksum University, Aksum, Ethiopia

[2]Centre for Development Research, University of Bonn, Walter-Flex-Street 3, 53113 Bonn, Germany

[3]International Centre for Tropical Agriculture (CIAT), Chitedze Agricultural Research Station, Lilongwe, Malawi

ABSTRACT

Even though scientific information on spatial distribution of hydrophysical parameters is critical for understanding erosion processes and designing suitable technologies, little is known in Geographical Information System (GIS) application in developing spatial hydrophysical data inputs and their application in Morgan-Morgan-Finney (MMF) erosion model. This study was aimed to derive spatial distribution of hydrophysical parameters and apply them in the Morgan-Morgan-Finney (MMF) model for estimating soil erosion in the Mai-Negus catchment, northern Ethiopia. Major data input for the model include climate, topography, land use, and soil data. This study demonstrated using MMF model that the rate of soil detachment varied from $<20\,t\,ha^{-1}\,y^{-1}$ to $>170\,t\,ha^{-1}\,y^{-1}$, whereas the soil transport capacity of overland flow (TC) ranged from $5\,t\,ha^{-1}\,y^{-1}$ to $>42\,t\,ha^{-1}\,y^{-1}$. The average soil loss estimated by TC using MMF model at catchment level was $26\,t\,ha^{-1}\,y^{-1}$. In most parts of the catchment ($>80\%$), the model predicted soil loss rates higher than the maximum tolerable rate ($18\,t\,ha^{-1}\,y^{-1}$) estimated for Ethiopia. Hence, introducing appropriate interventions based on the erosion severity predicted by MMF model in the catchment is crucial for sustainable natural resources management.

INTRODUCTION

Soil erosion is the dominant cause of soil degradation at a global scale [1–3]. This is accounted for between 70 and 90% of total soil degradation [2, 4]. The adverse influences of soil erosion as a cause for soil degradation have long been recognized as a severe problem for sustainability of economic development [5]. This is because a large portion of fertile soil is lost annually which negatively influences the goal of achieving food security [3]. However, estimation of soil erosion rate is often difficult due to a complex interplay of many factors, besides the differences in scale and methodological components of the studies [6, 7]. In a country like Ethiopia, with an agriculture-based economy for more than 85% of population, having reliable soil loss data is indeed a matter of great concerns and not a matter of choice. Regardless of the great deal of management practices undertaken aggressively in Ethiopia catchments by the government in the past 1-2 decades to reduce soil degradation, soil erosion by water is still recognized to be a severe threat to the national economy [2, 6, 7]. This indicated that the existing literature on the rate of soil erosion in Ethiopia calls for a wise decision supporting tools in order to reduce the degradation level. For instance, past studies on soil erosion in the catchments of Tigray region (northern Ethiopia) showed variability ranging from $7\,t\,ha^{-1}y^{-1}$ [8] to more than $24\,t\,ha^{-1}y^{-1}$ [7] and $80\,t\,ha^{-1}y^{-1}$ [9]. According to the report by FAO [10], erosion rate is estimated up to $130\,t\,ha^{-1}y^{-1}$ from cropland and $35\,t\,ha^{-1}y^{-1}$ averaged over all land use types in the highlands of Ethiopia. Such discrepancies in the rates of erosion by the studies mainly attributed to changes and differences in land use, management practices, and methods employed while developing input data and their respective scale of analysis. Predominantly, previous erosion related input data were developed from simple point observation such as runoff plot, and data were interpolated through conventional method [8, 11, 12]. Such method poses many limitations in terms of cost, representation, and reliability of the resulting data [5]. Recently, to reduce such limitations geostatistic techniques that interpolate data for an entire catchment from appropriately sampled point measurements are readily available [13, 14].

Mapping through conventional methods demands an intensive data collection, which is often difficult to practice in complex terrains like in northern Ethiopia [7, 14]. The Geographic Information System (GIS) techniques can provide easy and time effective tools to map and analyze erosion input data of hydrophysical parameters. These techniques coupled with the concept of catchment priority can help in identifying areas where treatment plans should be first located. Many studies (e.g., Sharad et al. [15]; Sanware et al. [16]) revealed that GIS techniques can have a great role in characterization and prioritization of subcatchments. The catchment level assessment and mapping of hydrophysical resources can support the identification of constraints, ecological problems, and adoption of effective management practices that sustain land and water resources using integrated catchment management strategies [17]. In addition, the availability of hydrophysical parameters in a GIS map format can be used readily for erosion model running in order to understand spatial distribution of ecological problems such as soil erosion. In many environmental studies, data inputs are measured at single points in space, even though classical statistics assume that measured data are independent and thus are not sufficient to analyze spatially dependent variables [13, 14, 18]. However, information is required for the entire catchment space, which necessitates methods that interpolate data to estimate the mean value within an area [13]. Geostatistics provides the basis for interpolation spatial variability of hydrophysical erosion model input parameters that affect runoff and soil loss [13, 14, 19–21].

To estimate soil erosion and suggest appropriate management plans, many erosion models such as Universal Soil Loss Equation (USLE) [22], Morgan-Morgan-Finney (MMF) [23], Water Erosion Prediction Project (WEPP) [24], Soil and Water Assessment Tool (SWAT) [25], European Soil Erosion Model (EUROSEM) [26], and Annualized Agricultural Non-Point Source (AnnAGNPS) [27] have been developed and used data inputs generated through GIS. Among these models, the USLE has remained the most practical method of estimating soil erosion potential for more than 40 years [28, 29], despite the fact that it has many limitations for application at catchment-scale. On the other-hand, process-based erosion models developed afterward have limitations in applicability due to intensive data and computation requirements [30]. The application of process-based models is not always an easy task since these require large amounts of information which is often not available, mainly in data scarce developing regions. The MMF model was selected to estimate annual soil loss, since this model endeavours to retain the simplicity of USLE and also encompasses the understanding of erosion processes into water and sediment phases [23]. Meaning, the MMF model was selected to be applied in this study because of its simplicity and flexibility as compared to the physical-base models and has a stronger physical base than USLE. In addition, since the MMF model is a physically based-empirical model (mix model), it needs less data than most of the other erosion predictive models [23].

Understanding the hydrophysical parameters that can influence erosion rate in a catchment is complex due to the combined nature of the natural processes and man-made features [7, 31]. Therefore, research to obtain quantitative description of hydrology/erosion in a catchment must consider these spatial heterogeneities. In order to tackle against hydrological related problems (runoff, sedimentation), accurate representation or locating the spatial distribution and variability of the influencing parameters using GIS is necessary [7, 32]. This study was aimed to derive and assess the spatial distribution of hydrophysical parameters developed using GIS technique and apply them in the MMF model for estimating the spatial variability of soil loss in the Mai-Negus catchment, northern Ethiopia. The spatial map can be used for prioritizing areas within the catchment that require immediate management measures on the basis of the severity of runoff/soil loss.

MATERIALS AND METHODS

Study Area

The study area, Mai-Negus catchment, is located in the Tigray region, northern Ethiopia (Figure 1). The catchment has an area of 1240 ha and altitude ranges from 2060 to 2650 m above sea level [33]. In the catchment, mean annual temperature of 22°C and annual rainfall of 700 mm have been recorded (Meteorology Agency-Mekelle branch). The highest amount of rainfall (>70%) is received between July and August. Land use is predominantly arable, with teff (Eragrostis tef) being the major crop along with different proportions of pasture land and scattered patches of trees, bushes, and shrubs. The major rock types are lava pyroclastic and metavolcanic. Leptosols are found mainly on the very steep positions, Cambisols on gentle to steep slopes, and Vertisols on the flat areas of the catchment [34].

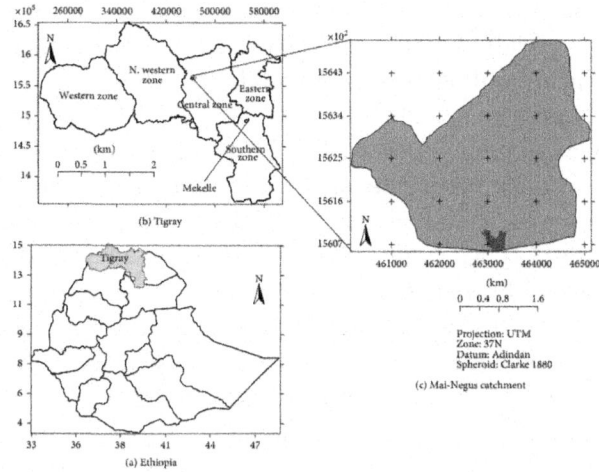

Figure 1: Study area: Ethiopia (a), Tigray (b), and Mai-Negus catchments (c). The blue colour shaded area is the reservoir.

Research Approach

The research approach in this study consisted of five main steps. These are (i) identification of hydrophysical parameters which are inputs of MMF model, (ii) field surveys and informal discussions in order to identify representative soil sampling zones within the study catchment for soil sampling and analysis and also the corresponding vegetation cover conditions, (iii) application of empirical relations which are described by Morgan et al. [23, 35] to calculate intermediate MMF model inputs, (iv) application of geostatistic interpolation technique for spatial model inputs development, and (v) application of MMF model in GIS environment while estimating spatially distributed erosion outputs such as total overland transport capacity and soil detachment rate.

MMF Model Inputs Preparation

Input data for MMF model include rainfall (mm), land use, digital elevation model (DEM) for slope map derivation, soil texture, soil moisture content at field capacity (%w w^{-1}), soil detachability index (g J^{-1}), bulk density of soil (Mg m^{-3}), cohesion of soil surface (KPa), soil moisture storage capacity (Rc), effective hydrological top soil depth (EHD), and ratio of actual to potential evapotranspiration (E_t/E_o) [23, 35]. Input data were collected from different sources such as field and laboratory determination, empirical relations, and the literature. Meteorological data such as rainfall and Et data were obtained from the meteorology station near the study area in 2009. Slope was derived from DEM developed from the topographic-map available at the Ethiopian Mapping Agency for Aksum area [33]. The map was scanned, and contours and spot heights were digitized and tagged with elevation values in a GIS environment. The vector elevation map was converted to raster and projected using the Universal Transverse Mercator 37 North (UTM-37N) reference system.

Crop and soil parameters were collected from 117 plots scattered throughout the catchment considering major land use and cover types (bush land, protected area, cultivated, abandoned fields, grazing land, mixed-forest, and residential). Supervised classification and visual interpretation of the land satellite image of November 2009 was carried out for general land use and cover mapping. In addition to this, crop covers for the different crop types and their corresponding geographic coordinates were collected using field survey in September 2009. Data related to rainfall such as rainfall intensity, number of rainy days, and total rainfall were assumed to be similar in the study catchment. The reason for having only one weather station in the study catchment is that the Office of Meteorology Agency believed that rainfall variability is negligible within such a small area regardless of the differences in elevation. Rainfall intensity was assumed at $25\,mm\,h^{-1}$ which is erosive for tropical climates such as Mai-Neguse catchment because no actual intensity data was found for the study catchment. Soil detachability index $(K)\,(g\,J^{-1})$ was determined from the literature that corresponds to the soil texture observed in the study catchment.

Soil Sampling Zones and Sample Collection

In order to prepare MMF model soil related inputs, soil sampling that considered soil variability in the study catchment was executed. Sampling approaches that divided a field into small units (zone sampling) can capture variability and provide more information about soil-test levels compared with one composite sample collected from an entire large sampling area [36]. To reduce the number of samples and sampling costs zone sampling is suggested to provide a way to group the spatial variability of soils while maintaining acceptable information about soil properties [36]. Sampling by zone assumes that sampling areas are likely to remain temporally stable [36, 37].

In this study, the zone sampling technique (divide a field into homogenous units that allow capturing variability and provide more information) was used to collect soil samples based on previous and existing knowledge of the soil and land use systems in the entire study catchment. The natural and management factors across the landscape that influenced soil properties spatial variability were considered while identifying the soil sampling zones. Three soil sampling zones that represented the soil quality (SQ) categories, long-term land use and soil management systems, and different erosion status sites in the catchment were identified using farmers' opinions and researcher and extension experts' judgment. The data that divided the catchment into the soil sampling zones was derived during the field reconnaissance surveys in June 2009. The SQ sampling zone was entirely used for arable land in the catchment whereas the other two sampling zones belonged to all the land use systems in the catchment. The sampling zones were further subdivided into different subsampling zones considering the variability within each zone and analytical costs.

The SQ sampling zone was divided into three subzones as high, medium, and low SQ based on farmers' knowledge. They used indicators such as yield and yield component, soil depth, colour, and fertility conditions to divide into these subzones. The details on how local farmers' classified soil into different SQ categories in the study catchment can be found in Tesfahunegn et al. [38].

Eight representative long-term land use system sampling zones were identified based on farmers' historical and present information acquired in the catchment. These are (i) natural forest; (ii) plantation of protected area;

(iii) grazed land; (iv) teff (Eragrostis tef)-faba bean (Vicia faba) rotation; (v) teff-wheat (Triticum vulgare)/barley(Hordeum vulgare) rotation; (vi) teff monocropping; (vii) maize (Zea mays) monocropping; and (vii) uncultivated marginal land. The age of the systems varied from 5-6 years for teff monocropping and 20–30 years of maize monocropping system. Average age of the other systems was about 10 years except for the plantation, grazed land, and uncultivated marginal land systems with more than 15 years.

The erosion status-based sampling zone was divided into three subzones as stable, eroded, and deposition (aggrading) sites. Information from the local farmers, extension agents, and researcher's (first author) observation on the level of topsoil depth (A-horizon), deposition, rills, pedestals, root and subsoil exposure, and gullies indicators were considered while identifying the three erosion-status sampling subzones. Those areas having A-horizon and minimum erosion indicators were considered as stable sites and the reverse of this as eroded sites. Depositional sites were also easily identified as they are mainly located in depression and flat areas with evidences of recent sediment deposition. In total, there were 14 subsampling zones across the erosion-status sites in the catchment for the soil samples collection. After doing all this identification and division, the soil sampling points in each subzone were located at the centre, considering soils in that point best represent the samples. Each sampling point was georeferenced as their distribution in the catchment is shown in Figure 2. The sampling distance was not regular, ranging from 40 to 180 m.

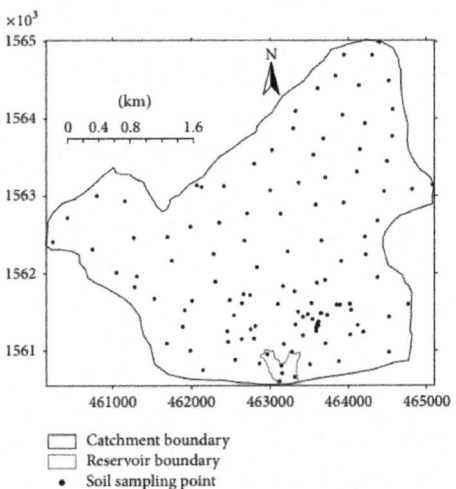

Figure 2: The distribution of soil sampling and vegetation cover points in the study catchment.

Soil samples were collected in June 2009. A total of 51 soil samples (3 subzones × 17 samples) were collected from the SQ based sampling zone. From the long-term land use systems, a total of 24 soil samples (8 subzones × 3 samples) were collected. It was also collected 42 soil/sediment samples (3 subzones × 12 samples in the catchment and 6 sampling points in the reservoir) from the erosion-status sites. The grand total of the composite samples collected across the sampling subzones was 117. Each composite soil sample was collected using 5–8 samples from each representative subsampling zone depending on the size and homogeneity of the sampling area (100–300 m²). All the composite soil samples were collected at the soil depth of 0–20 cm (the plough depth) since this is where most changes are expected to occur due to erosion, long-term land use, and soil management practices. The composite soil samples were pooled into a bucket and mixed thoroughly to homogenize it. Finally, a subsample of 500 g from the pooled composite samples was taken and soil samples were air dried and sieved to pass 2 mm mesh sieves before analysis for soil textures. On the other hand, two undisturbed soil samples were collected from each soil sampling point for bulk density and soil moisture determination. In addition, field level observation and measurement for parameters such as effective hydrological top soil depth (m), ground cover, and cover factor were carried out from the sampling points and georeferenced.

Soil Analysis

The soil samples collected in the soil sampling zones were determined for soil texture using the Bouyoucos hydrometer method [39], soil bulk density (BD) by the core method [40], and soil moisture content at field capacity (w w^{-1}) by equilibrating the soil with water through capillary action in KR box [41].

Empirical Relations in Deriving Inputs of MMF Model

Some intermediate input parameters were estimated from observed data in the catchment using the empirical relations described in Morgan et al. [23] as

$$E = R\left(11.9 + 8.7\log_{10}(I)\right),$$

$$R_c = 1000 {*} MS {*} BD {*} EHD {*} \left(\frac{E_t}{E_o}\right)^{0.5},$$

$$SR = R\exp\left(-\frac{R_C}{R_o}\right),$$

$$R_o = \frac{R}{R_n},$$

$$(1)$$

where E is annual kinetic energy of rainfall (J m^{-2}), I is intensity of rainfall which is assumed to be 25 mm h^{-1} in tropical conditions, SR is surface runoff/ overland flow (mm), R_n is number of rainy days, R is average annual rainfall (mm), R_c is soil moisture storage capacity (mm), R_o is annual rain per rain day, MS is soil moisture content at field capacity (w w^{-1}), BD is bulk density of the topsoil layer (Mg m^{-3}), EHD (m) is effective hydrological topsoil depth defined as the depth of soil from the surface to an impermeable or stony layer to the base of A horizon or to the dominant root base, and E_t/E_o is the ratio of actual (E_t) to potential (E_o) evapotranspiration. EHD is the top soil depth within which the storage of water affects the generation of runoff.

Intermediate maps derived on the basis of land use/cover map (Figure 3) included ratio of actual to potential evapotranspiration (E_t/E_o), permanent rainfall contributing to permanent interception, and stream flow (A) and crop cover management factor (C_f). The C_f combines C and P factors of the Universal Soil Loss Equation to give ratio of soil loss under a given management to that of bare ground with down-slope tillage, other conditions being equal. These were determined in the field. Intermediate layers derived from soil map (soil texture) included soil detachability index (K) and cohesion of topsoil (COH) that were generated using ArcGIS 9.3 software. According to Morgan et al. [23] and Dinka [35], K is defined as the weight of soil detached from the soil mass per unit of rainfall energy. The values of plant and soil related hydrological parameters are shown in Tables 1 and 2, respectively. Inputs such as plant related (e.g., EHD, A, CC) and soil related (e.g., K, COH) parameters were adopted from Morgan et al. [23] and Dinka [35], in which such values corresponded to crop type and cover conditions and soil textures observed in the field.

Table 1: Plant related hydrological parametersfor different land use and land cover in the Mai-Negus catchment

Land use	EHD (cm)	A	CC (%)	PH (m)	GC (—)	E_t/E_o	C_f
Teff	5	0.15	85	0.2	0.85	0.5	0.5
Barley/wheat	12	0.3	60	0.3	0.80	0.58	0.3
Maize	12	0.25	50	1	0.75	0.68	0.25
Pulse	12	0.2	55	0.2	0.80	0.65	0.25
Grazing	12	0.2	80	0.18	0.80	0.8	0.1
Closure (grass/shrubs/bushes)	15	0.4	55	1.5	0.60	0.85	0.1
Mixed forest	20	0.5	65	1.8	0.65	0.9	0.01
Residential	19	0.1	5	0.75	0.55	0.6	0.13

EHD: Effective root depth; A: the percentage of rainfall contributing to permanent interception; CC: canopy cover fraction; PH: plant height; GC: ground cover; E_t/E_o: the ratio of actual to potential evapotranspiration; and C_f: the crop cover management factor.

Table 2: Soil related hydrological parameters values for soil texture in the Mai-Negus catchment

Texture	K (gm J^{-1})	MS (% w w^{-1})	BD (Mg m^{-3})	COH (KPa)
Sandy loam	0.7	18.6	1.87	2
Sandy clay loam	0.3	23.7	1.72	3
Clay loam	0.4	35.38	1.48	10
Sandy clay	0.35	21.21	1.51	1
Silt clay loam	0.3	28.63	1.31	9

K: soil detachability index; MS: soil moisture at field capacity (1/3 bar) tension; BD: bulk density top soil; and COH: cohesion of topsoil.

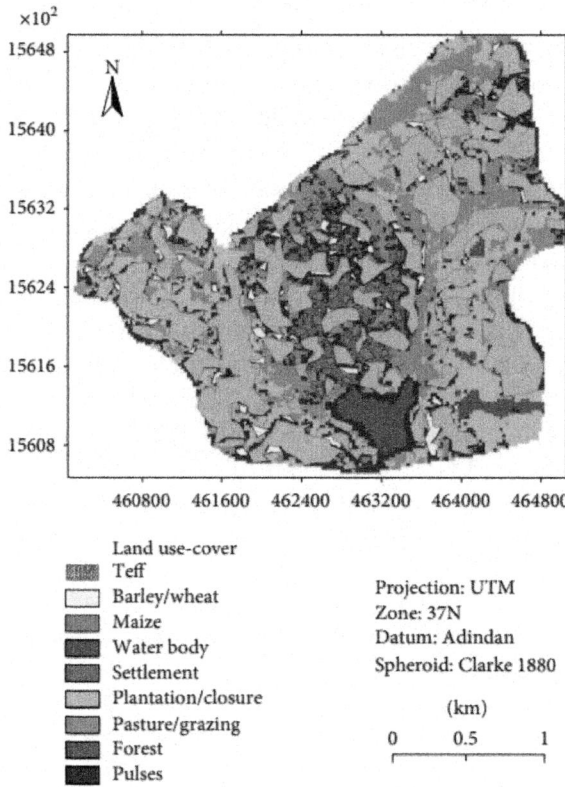

Figure 3: Land use and land cover of the study catchment.

Application of Geostatistical Interpolation Technique

After the point data and their corresponding coordinates were entered into ArcGIS 9.3 software, maps of hydrophysical model input parameters were developed using kriging interpolation technique [42]. Ordinary kriging was selected as the preferred interpolation method for MMF model spatial inputs derivation because it was more reliable than the other interpolation methods based on the mean squared error which compares the measured values with the predicted ones. Moreover, since the spacing of the measured or observed hydro-physical input parameters were relatively sparse and randomly chosen for each subsampling zone, ordinary kriging is the best unbiased predictor for the random process at specific unsampled locations [43]. Ordinary kriging also has an additional advantage of minimizing the influence of outliers [44]. The semivariogram analyses were conducted before the application of ordinary kriging interpolation of the input parameters. This is because the semivariogram model determined the interpolation function [14]. Semivariogram models were chosen by using the cross-validation technique that compares statistical mean square error values estimated from the semivariogram models and actual values.

MMF Model Application

The MMF modelling processes erosion in two phases, that is, the water and sediment phases [23]. The water phase mainly comprises of prediction of soil detachment by rain splash. It thus requires data related to intensity of rainfall (I, mm h^{-1}), number of rainy days (R_n), and average annual rainfall (R, mm). After developing the different input spatial maps (layers), the rate of soil detachment by rain drop impact (F, kg m^{-2}), rate of soil detachment by runoff (H, kg m^{-2}), and transport capacity of overland flow (runoff) (TC, kg m^{-2}) are calculated in the GIS environment as follows:

$$F = 10^{-3} * K * \left(E * e^{-0.05A}\right),$$

$$H = 10^{-3} * (0.5\text{COH})^{-1}(\text{SR})^{1.5}\sin(S)(1 - \text{GC}),$$

$$\text{TC} = 10^{-3} * C_f * \text{SR}^2 * \sin(S),$$

$$(2)$$

where K is soil detachability index (g J^{-1}), E is annual kinetic energy of rainfall (J m^{-1}), A is percentage of rainfall contributing to permanent interception and stream flow (%), COH is cohesion of the soil surface (KPa), GC is fraction of ground (vegetation) cover (0^{-1}), Cf is the crop cover management factor, and S is the steepness of the ground slope expressed in degree. Total particle detachment ($D = F+H$) is finally computed as sum of soil particle detachment by runoff (H) and soil particle detachment by raindrop (F) impacts. The model compares the predicted rate of splash detachment (D) and the transport capacity for overland flow (TC), and the minimum value is taken as the erosion rate (annual soil loss) estimated for fthe study catchment (Figure 4).

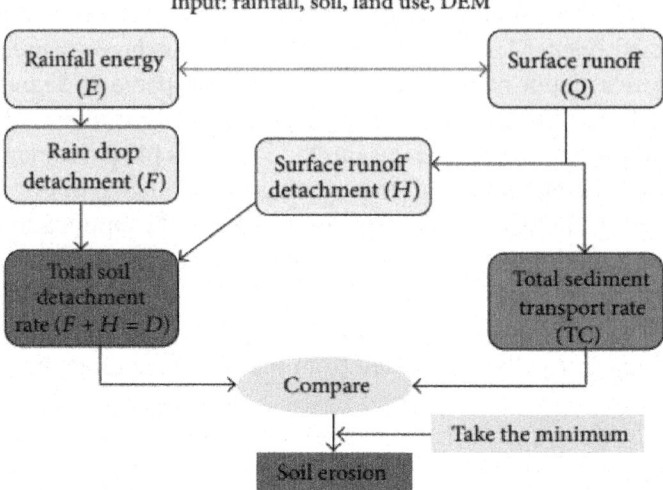

Figure 4: Flow chart depicting the methodology used for MMF modelling (source: Morgan et al. [23]).

Model Evaluation

The percent difference (D) was used as methods for goodness-of-fit measure of model prediction. The model estimated soil loss rate in this study was evaluated with respect to the sediment deposited surveyed in the reservoir. The percent difference (D) measures the average difference between the simulated and measured values as

$$D = \left(\frac{p - q}{q} \right) 100,$$

(3)

where p is model simulated value and q is measure value. "*A* value close to 0% is best for D; however, higher values of D are acceptable if the accuracy in which the observed data have gathered is relatively poor" [45].

In addition, different studies showed that soil profile data such as degree of truncation of the top soil horizon by erosion can be used to assess the performance of models (e.g., Tamene [7]; Desmet and Govers [46]; Mitasova et al. [47]; and Turnage et al. [48]). In this study, the soil profile data was applied to evaluate the estimated erosion result by MMF model. For this purpose, areas with possible erosion processes were selected, and soil profile data related to the truncation level of the A horizon were documented in the study catchment. These data were then compared with the soil loss prediction made by the model. The main purpose was to evaluate whether the spatial pattern of erosion predicted by MMF model correlated well with the depth of soil profile data which semiquantitatively verified the performance of the model. With regard to this, 10 soil profiles (pits) were identified and georeferenced and then compared to the corresponding model spatial erosion results. The MMF model result also compared with the outputs of other models which simulated in the study area and other highlands of Tigray region in northern Ethiopia.

RESULTS AND DISCUSSION

Spatial Distribution of Hydrophysical Parameters Influencing Erosion

Spatial variability in slope, rainfall, vegetation, soil texture, and land use and cover are among the main factors which influence the distribution of erosion risk in a catchment [7]. Since there was only one meteorological station, weather data such as rainfall was assumed to be the same throughout the study catchment. The variability in the kriging interpolated maps of the other erosion influencing factors in the catchment is shown in Figure 5. The slope of the study catchment (Figure 5(a)) showed that the northern and eastern part of the catchment had a very steep slope (up to 77°) and slope steepness decreases in the direction to the reservoir (south of the catchment). This indicated that sources of hydrological losses as runoff and sediment yield can be higher on steep areas as compared to flat to gentle slopes provided that the other factors are similar. Such influence of slope can be explained by soil texture variability; that is, silt and clay soils dominated towards the reservoir and flat areas whereas there was a coarser texture on the steep slopes of the catchment (Figure 5(b)). This could be associated with the selective behaviour of erosion in transporting fine particles [7, 49].

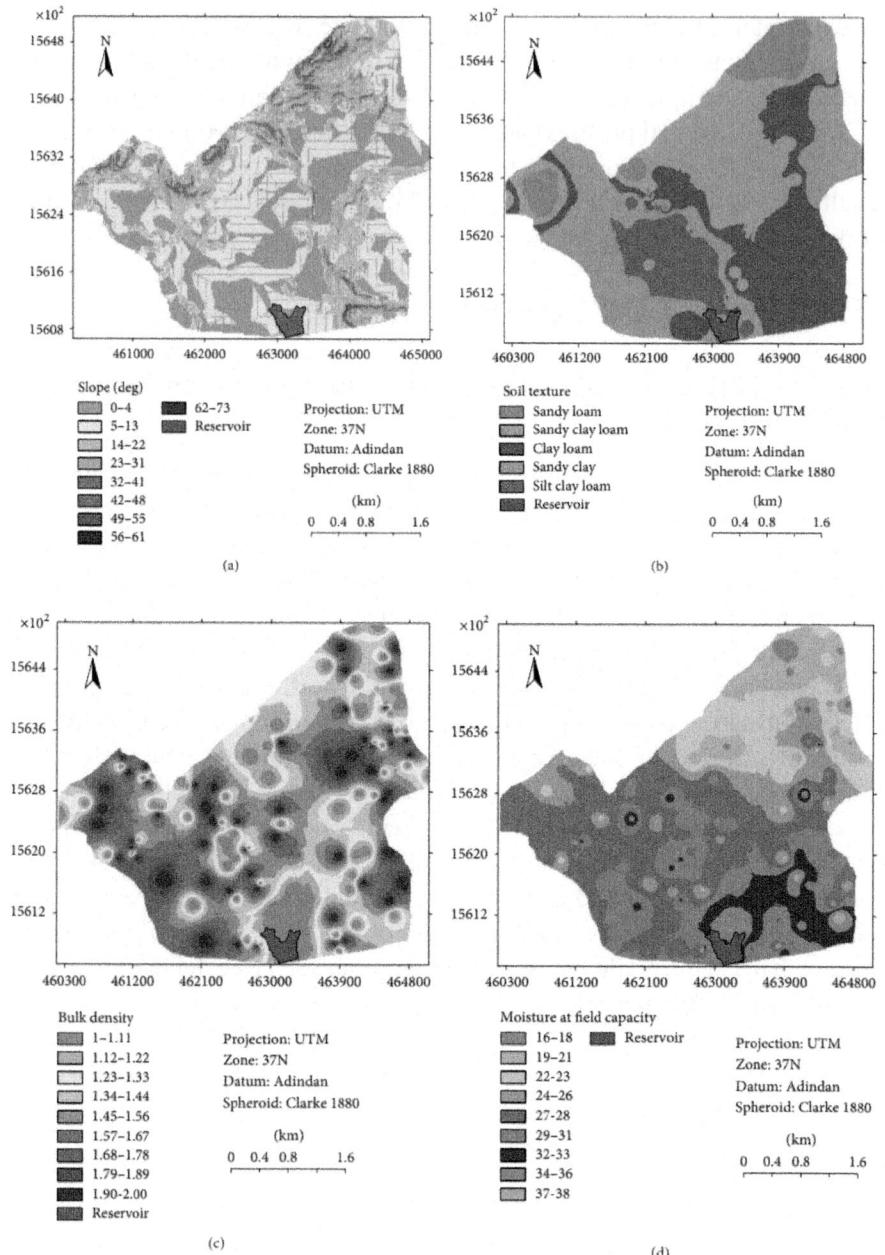

(a)

(b)

(c)

(d)

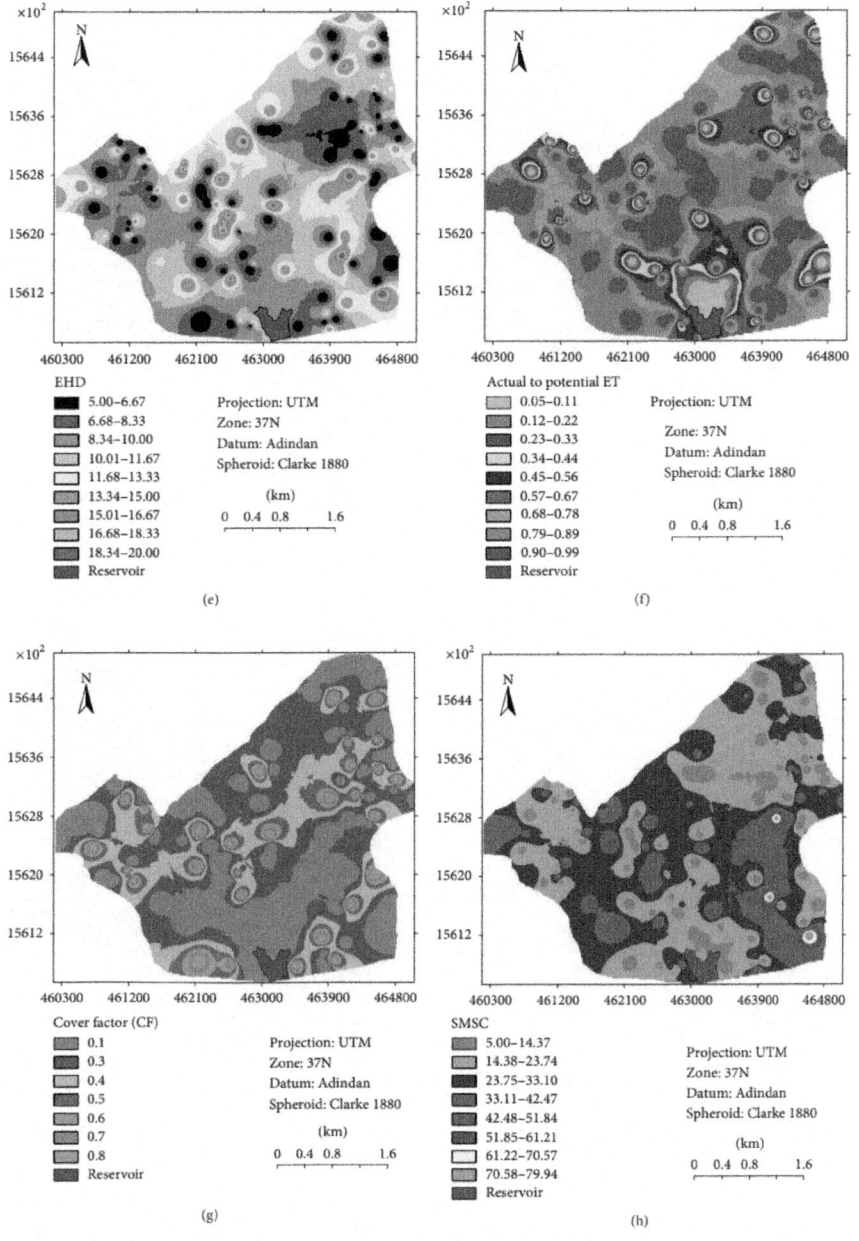

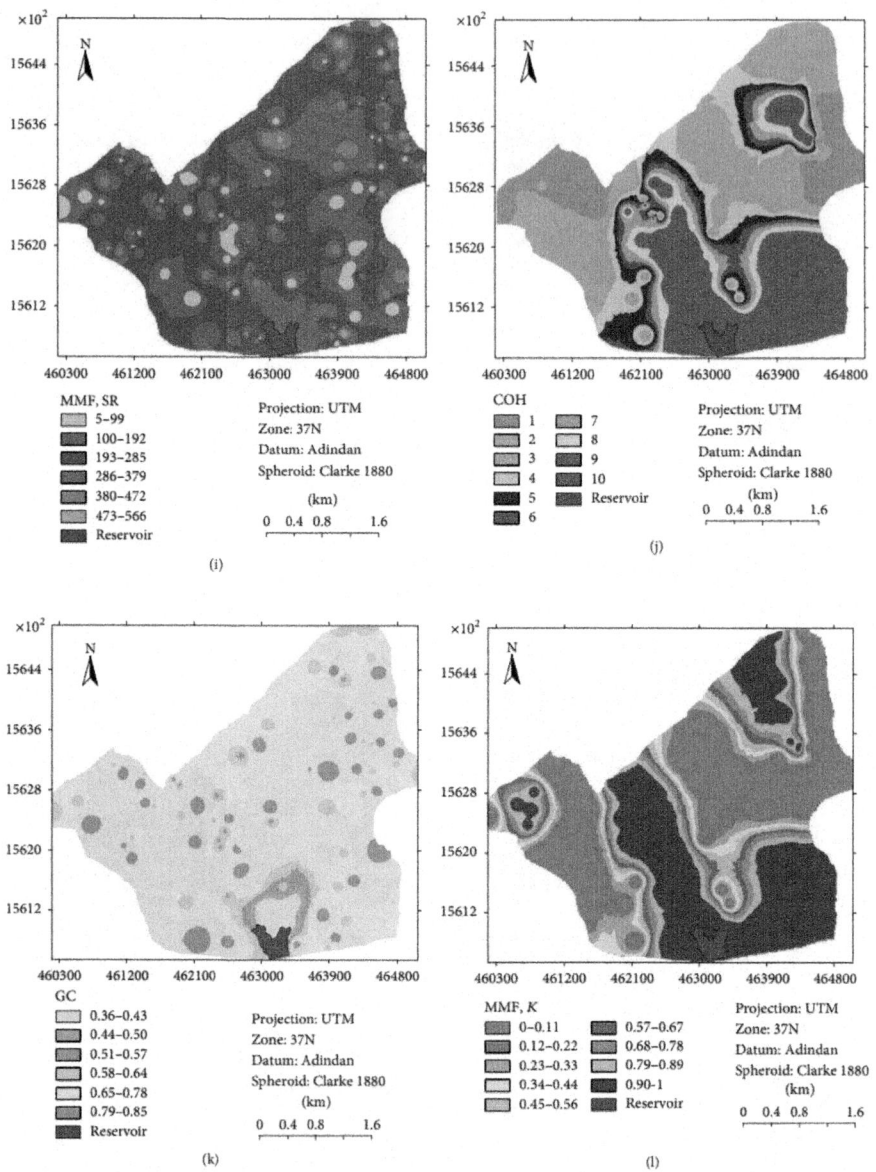

Figure 5: Spatial distribution of parameters influencing the hydrology/erosion of the catchment: (a) slope; (b) soil texture; (c) bulk density (Mg m^{-3}); (d) moisture at field capacity (%); (e) EHD, effective hydrological top soil depth (cm); (f) E_t/E_o, actual to potential evaporation; (g) cover management factor; (h) SMSC, soil moisture storage capacity (mm); (i) SR, surface runoff (Q) (mm); (j) COH, cohesion of the soil surface (KPa); (k) GC, fraction of ground cover (0-1); and (l) K, soil detachability index.

Bulk density is higher (up to $1.90 \, \mathrm{Mg \, m^{-3}}$) in the steep slopes with poor vegetation cover and soil management practices and decreased to about $1.10 \, \mathrm{Mg \, m^{-3}}$ around the reservoir and valley which are located at the foot-slope of the catchment (Figure 5(c)). From this figure, it is possible to observe higher erosion on areas dominated by higher bulk density as they are located on the steep slopes, low SQ, eroded sites, and marginal land soils and soils with poor vegetation cover. An increase in bulk density can negatively affect the circulation of air, water, and plant nutrients and their root system and in-turn raises rate of soil erosion [41, 50]. Such higher BD increases surface runoff which is the driving force for soil loss by decreasing soil infiltration and soil water holding capacity [51].

The spatial distribution of soil moisture content at field capacity (MS) showed lower value in the hilly part of catchment as compared to that of flat to gentle slopes under similar management and cover conditions (Figure5(d)). This is consistent with the report by Behera et al. [52] who reported lower values of soil parameters such as MS in the topsoil layer of hilly areas. However, regardless of the slope steepness the highest MS was found in some sites of the study catchment such as in the forest and closed pasture lands with relatively higher organic matter (data not shown). The lowest MS in the catchment was associated with low SQ, poor land use, and soil management systems (e.g., marginal land, over grazing land, and eroded sites with shallow soil depth).

The effective hydrological top soil depth (EHD) spatial map (Figure 5(e)) indicated higher values (>15 cm) in relatively better vegetation-cover areas, high SQ and stable sites, and flat to gentle slopes. The EHD values were lower (15 cm) in relatively better vegetation-cover areas, high SQ and stable sites, and flat to gentle slopes. The EHD values were lower (15 cm) in marginalized areas, cultivated and degraded grazing lands. Majority (78%) of the study catchment showed low EHD, indicating that such sites can be the source of higher runoff and soil loss. Similarly, the higher values (0.68 to 0.90) of E_t/E_o corresponded to forest land, whereas lower values ($0.05-0.33$) corresponded to agricultural areas. The intermediate E_t/E_o values can correspond to other areas such as bush land (Figure 5(f)). Such values are thus more influenced by the spatial distribution of crop cover and management factor (C_f) which ranges from $0.2-0.8$ (Figure 5(g)). Areas of the catchment with higher C_f are expected to have lower E_t/E_o values and vice versa [52]. As part of the C_f factor, vegetation maintains high rates of evapotranspiration, rainfall interception, and runoff infiltration [52, 53].

The soil moisture storage capacity (R_c) calculated as a function of MS, BD, EHD, and E_t/E_o varied spatially from 5 to 79 mm (Figure 5(h)). Most of the small R_c values were located on steep slopes with shallow soil depth, poor surface cover, and marginal lands. Farmers who cultivated their land on steep slopes confirmed that they often face crop failure due to moisture stress related to low R_c. Generally, low R_c can be used as an indicator of a source of higher runoff. Such soil could be susceptible to soil detachment by raindrop and runoff impacts as a result of less vegetation cover. Reduction in the EHD (Figure 5(e)) can lead to low soil moisture storage capacity, R_c (Figure 5(h)), which resulted in higher surface runoff (Figure 5(i)).

For estimating soil detachment rate by runoff (H), understanding the spatial distribution of slope (S) (Figure 5(a)), surface runoff (SR) (Figure 5(i)), cohesion of the soil surface (COH) (Figure 5(j)), and fraction of ground cover (GC) (Figure 5(k)) are important conditions. The soil detachability by raindrop impact is also influenced by the soil detachability index (K) which showed higher values (0.79–0.98 g J−1) on flat areas with coarser soils and poor soil cover and lower values (0.11–0.33 g J−1) on steep slope with fine soils and good vegetation cover in the study catchment (Figure 5(l)). The lower K values on steep slope could be associated with rainfall drop impact angle which is falling to the ground surface on inclination.

Estimated Soil Loss Using MMF Model

The spatial distribution of the rate of soil detachment by raindrop (F) indicated that flat areas (south of the catchment) were exposed more to F as compared to that of hilly land (Figure 6(a)). This could be attributed to the perpendicular fall of raindrop energy (strong energy) on flat areas as compared to raindrop impact angle on inclination [2, 3]. This implied that steep slope reduces the impact of rainfall drop energy because rain drop met the soil surface on inclination. Despite this, the net detached and transported soil is almost the same on flat lands, whereas on steep surfaces more soil particles are thrown downslope than upslope during the detachment process, resulting in a net movement of material downslope [3]. Generally, the F values estimated by the model ranged from 160 t ha^{-1} y^{-1} in the study catchment. The plateau and valley parts of the catchment showed higher F values. Increasing soil cover and practicing zero grazing, for example, can be part of the solution for reducing the amount of soil detached by raindrops. This is because such practices can increase vegetation cover that dissipates rainfall and runoff energy [2, 3].

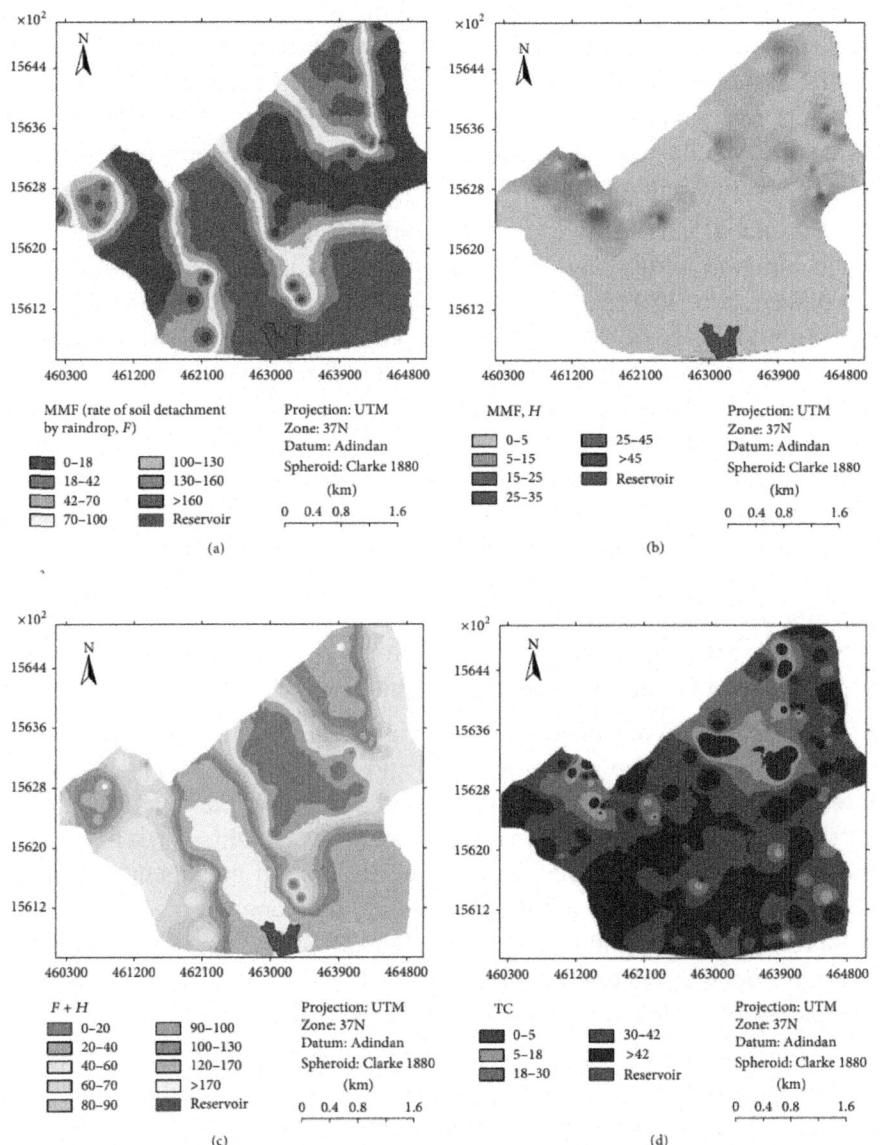

Figure 6: Spatial distribution of MMF erosion estimation. (a) Rate of soil detachment by rain drop impact (F) (t ha−1 y−1); (b) rate of soil detachment by runoff impact (H) (t ha−1 y−1); (c) total soil detachment ($F+H$) (t ha−1 y−1); and (d) soil transport capacity of overland flow (TC) (t ha−1 y−1).

The soil detachment rate by runoff (H) is higher ($35\,\mathrm{t\,ha^{-1}\,y^{-1}}$) on the steep-hilly parts of the catchment (Figure 6(b)), as slope steepness increases the amount and velocity of overland flow or surface runoff (SR) (Figure 5(i)). In addition, the expansion of cropland and open grazing practices to the steep parts of the catchment could make the soil more vulnerable to soil erosion by runoff. A similar view to this study was reported by Tamene [7] who stated that on steeper slopes the soils are likely to be thinner and the flow velocity is high, which results in high runoff; whereas runoff threshold is high since water is likely to pond and infiltrate, resulting in little or no runoff on flat slope land.

The spatial variability of the total soil detachment rate ($F+H$) as a result of the summation of F and H for the catchment is shown in Figure 6(c) and this ranged from 170 t ha^{-1} y^{-1}. The highest rates of $F+H$ occurred in low SQ fields, marginal lands, and subsoil exposed soils having low soil resistance to detaching forces. The lowest was observed in forest land, protected plantation areas, and farm lands with high soil quality regardless of the slope steepness. This study generalized that the rate of soil loss increased with an increase of detaching forces. It was observed from the field that the process of erosion can continue until first the topsoil and finally the subsoil disappear unless suitable controlling measures are implemented. However, it is a matter of fact that all soils detached cannot be reached at the outlet of the catchment because of deposition areas on the way to the outlet. It is therefore important to have information that shows the spatial distribution of soil transport capacity of the overland flow (TC) in the study catchment.

The spatial distribution of TC (Figure 6(d)) indicated higher values (>42 t ha^{-1} y^{-1}) on steep slope, marginal and over grazed lands, and sites with bare soils and intensively cultivated without proper soil management and conservation measures, whereas it indicated lower values (<5 t ha^{-1} y^{-1}) from relatively less disturbed areas (better vegetation and management practices). This indicated that TC is influenced not only by slope but also by cover crop, supporting practices, and soil erodibility and erosivity conditions. These could be the factors that led to the irregular spatial distribution and variability of the soil loss as TC which was estimated by MMF model.

In general, the TC value was lower than that of $F + H$ which attributed to the transport of fewer amounts of detached soils by rainfall drop and runoff impacts. Thus, erosion is transport limited in the study catchment. However, there were conditions whereby TC could be larger than $F+H$, for example, steep areas, compact soils, marginal lands, and farmlands with poor SQ. This could be related to the fact that in steep slope land, the possibility of deposition to take place in the natural flow is low and the time for soil infiltration is also short. The mean and spatial distribution of total soil detachment ($F+H$) was higher than the TC, indicating that the value of TC is taken to show the magnitude of annual soil loss from the study catchment. Thus, considering the TC values, the higher erosion risk areas can be identified scattered throughout the catchment. The model generally predicted erosion being limited by transportation, indicating that the amount of soil detachment is very high in the study catchment. This indicated that soil cover is too poor in which soil is exposed to detaching forces and also, the steep slope which generate high runoff can detach and transport large mass of soils.

Results of MMF Model Related to Reservoir Sediment Survey and Soil Profile Data

The model estimated soil loss at catchment level was compared with the survey based measured sediment yield from the reservoir located at the outlet of the catchment. The average catchment level soil loss (TC) estimated by MMF model was $26\,t\,ha^{-1}y^{-1}$ whereas the sediment yield measured from the reservoir survey was $20\,t\,ha^{-1}y^{-1}$. This resulted in percent difference (D) value of 30%. The similarity between the measured and model simulated value at catchment level was 70%, which is moderately acceptable. The MMF model seems to be overestimating the average soil loss (TC) from the entire catchment as compared to the sediment yield measured from reservoir survey, indicating that the model can pronounce erosion processes observed in the catchment using some extremely higher values of soil loss rate.

In addition, the soil erosion rate predicted by the model was compared with observed soil profile depth data in selected sites of the study catchment. The MMF model was assessed in terms of its capacity to identify areas of soil truncation and/or to predict lower erosion on areas where buried soils and/or alluvial/colluvial deposits or stable soils were observed. The MMF model accurately predicted erosion in about 80% of the pits observed in the catchment whereas the 20% disagreement was located at the upslope position of the catchment where the model predicted slight erosion for sites of a very truncated soil profile. In most sites (80%) of the catchment that characterized by truncated soil profile the model estimated high soil loss rate ($>20\,t\,ha^{-1}y^{-1}$).

MMF Model Evaluation in relation to Other Studies (Models)

For most of the catchment (>80%), the MMF model predicted higher soil loss rates than the maximum tolerable soil loss rate of $18\,t\,ha^{-1}y^{-1}$ estimated for the country [11]. This could be related to the reason that the study catchment could be characterized by very low soil infiltration, low soil water holding capacity and poor vegetation cover (inappropriate land use system), and low conservation measures. Consistent with this finding and explanation many previous reports indicated that a higher soil loss rate is strongly associated with high runoff resulting from absence of runoff flow obstacles such as vegetation cover, conservation measures, impoundments, and soil with low infiltration rate and low soil water holding capacity [54–60]. If an average annual soil generation rate of $6 \, t \, ha^{-1} y^{-1}$ [61] is considered, the soil loss rates estimated by the model in most parts of the catchment could be beyond this acceptable level. It was less than 1% of the catchment area that experienced a soil erosion rate below $6 \, t \, ha^{-1} y^{-1}$.

This study showed lower sediment yield ($26 \, t \, ha^{-1} y^{-1}$) as compared to a previous study using Soil and Water Analysis Tool (SWAT) model simulated result ($30 \, t \, ha^{-1} y^{-1}$) in the same catchment [31]. However, such differences can be attributed to model variability in the scale of data requirement. In addition, SWAT model is a continuous and long-time simulation model that might not be represented by the existing land use which can increase model prediction uncertainty [31]. Generally, the variability in soil loss predicted by both models is not higher than 15%, which is an acceptable model uncertainty. Sediment yield of $21 \, t \, ha^{-1} y^{-1}$ based on an in-filled dam with a catchment area of $6.7 \, km^2$ in one of the Tekezze River's tributaries in the Tigray region is also reported by Machado et al. [62].

In addition, a previous study using four catchments in the Tigray region reported an average soil erosion rate of $24 \, t \, ha^{-1} y^{-1}$ [7]. In general, this and the above results are consistent with the trend of erosion estimated using MMF model in this study. Such comparison considers their similarity in the scale of measurement, input data requirement, and the farming system within the catchments. Therefore, the application of MMF model for estimating and identifying the severity of water erosion in order to introduce appropriate interventions is acceptable in conditions similar to the Mai-Negus catchment (study area).

CONCLUSION

This study concluded that GIS is a useful tool to integrate and manage spatially distributed hydrophysical data while assessing the spatial distribution of erosion. In this study, the MMF model results showed a lower rate of erosion for the soil transport capacity of overland flow (TC) when compared to the rate of soil detachment. This indicated that erosion is transport limited, and thus TC can show a realistic image of soil erosion hotspot sites in the study catchment conditions while introducing suitable soil conservation and/or management practices. Availability of such a study can help in making a quick assessment of runoff and soil loss status for the process of appropriate decision making. The average catchment level soil erosion rate (TC) estimated by MMF model was $26 \, t \, ha^{-1} y^{-1}$ but there were sites with erosion rates higher than $42 \, t \, ha^{-1} y^{-1}$. The sources of such higher soil loss rates were identified mainly to be the eroded sites, low soil quality soils, marginal land, overgrazed lands, and mono-cropping cultivated land system with poor soil management and conservation measures located in the mountainous (north-west) and central-ridge landforms of the study catchment. Therefore, introducing appropriate site specific interventions such as agroforestry, agronomic practices, exclosure of degraded lands, conservation measures based on the model erosion maps produced for the study catchment are suggested to be a practical solution for attaining sustainable environmental management and production services. However, the high soil detachment rates that occurred in many fields of the study catchment as a result of differences in spatial distribution of soil erosion influencing factors should be considered while designing and promoting appropriate practices to improve and maintain soil and water resources.

ACKNOWLEDGMENTS

The authors gratefully acknowledge the financial support of DAAD/GIZ (Germany) through the Centre for Development Research (ZEF), University of Bonn (Germany), and the support of Aksum University (Ethiopia) during the field work. The authors also greatly appreciate the assistance offered by the local farmers and extension workers during the field study.

REFERENCES

1.   S. J. Scherr, "Soil degradation: a threat to developing-country food security by 2020?" Food, Agriculture, and Environment Discussion Paper 27, International Food Policy Research Institute, Washington, DC, USA, 1999.

2.   R. Lal, "Soil degradation by erosion," Land Degradation and Development, vol. 12, no. 6, pp. 519–539, 2001.

3.   R. P. C. Morgan, Soil Erosion and Conservation, Blackwell, Malden, Mass, USA, 3rd edition, 2005.

4.   M. A. Zoebisch and E. DePauw, "Degradation and food security on a global scale," in Encyclopedia of Soil Science, R. Lal, Ed., pp. 281–286, Marcel Dekker, 2002.

5.   R. Lal, "Soil quality and sustainability," in Methods for Assessment of Soil Degradation, R. Lal, W. H. Blum, C. Valentine, and B. A. Stewart, Eds., pp. 17–30, CRC Press, Boca Raton, Fla, USA, 1998.

6.   J. Ananda and G. Herath, "Soil erosion in developing countries: a socio-economic appraisal," Journal of Environmental Management, vol. 68, no. 4, pp. 343–353, 2003.·

7.   L. Tamene, Reservoir siltation in the drylands of northern Ethiopia: causes, source areas and management options [Ph.D. thesis], University of Bonn, Bonn, Germany, 2005.

8.   J. Nyssen, Erosion processes and soil conservation in a tropical mountain catchment under threat of anthropogenic desertification—a case study from Northern Ethiopia [Ph.D. thesis], Katholieke University, Leuven, Belgium, 2001.

9.   G. Tekeste and D. S. Paul, "Soil and water conservation in Tigray, Ethiopia," Report of a Consultancy Visit to Tigray, University of Wageningen, Wageningen, The Netherlands, 1989.

10. FAO, "Ethiopian highland reclamation study: Ethiopia," Final Report, FAO, Rome, Italy, 1986.

11. H. Hurni, "Erosion—productivity—conservation systems in Ethiopia," in Soil Conservation and Productivity, Proceeding of the 4th International Conference on Soil Conservation, I. P. Sentis, Ed., pp. 654–674, Maracay, Venezuela, November 1985.

12. G. B. Tesfahunegn, Soil erosion modeling and soil quality evaluation for catchment management strategies in northern Ethiopia [Ph.D. thesis], University of Bonn, Bonn, Germany, 2011.

13. H. Pohlmann, "Geostatistical modelling of environmental data," Catena, vol. 20, no. 1-2, pp. 191–198, 1993.

14. G. B. Tesfahunegn, L. Tamene, and P. L. G. Vlek, "Catchment-scale spatial variability of soil properties and implications on site-specific soil management in northern Ethiopia," Soil and Tillage Research, vol. 117, pp. 124–139, 2011.

15. D. Sharad, M. V. Ravi Kumar, L. Venkataratnam, and T. Mel lerwara Rao, "Watershed prioritization for soil conservation- a GIS approach," GeoCarto International, vol. 1, no. 2, pp. 27–34, 1993.

16. P. G. Sanware, C. P. Singb, and R. L. Karale, "Remote sensing application for prioritization of subwatsheds using sediment yield and runoff indices in the catchment of Marani barrage (Sahibi),"UNDP/ FAO Project 13, Remote Sensing Center, AIS LUS, Government of India, New Delhi, India, 1998.

17. D. K. Das, K. S. S. Sharma, and N. Kalra, "Education and training in remote sensing and GIS for sustainable agricultural development," in Proceeding of the 15th Asian Conference on Remote Sensing, pp. 1–6, Banagalore, India, 1994.

18. E. Özgöz, "Long term conventional tillage effect on spatial variability of

some soil physical properties,"Journal of Sustainable Agriculture, vol. 33, no. 2, pp. 142–160, 2009.

19. R. Webster, "Quantitative spatial analysis of soil in the field," in Advance In Soil Science, B. A. Stewart, Ed., vol. 3, pp. 1–70, Springer, New York, NY, USA, 1985.

20. R. Webster and M. A. Oliver, Statistical Methods in Soil and Land Resource Survey, Oxford University Press, Oxford, UK, 1990.

21. C. A. Cambardella, T. B. Moorman, J. M. Novak et al., "Field-scale variability of soil properties in central Iowa soils," Soil Science Society of America Journal, vol. 58, no. 5, pp. 1501–1511, 1994.

22. W. H. Wischmeier and D. D. Smith, Predicting Rainfall Erosion Losses: A Guide to Conservation Planning, vol. 537, US Department of Agriculture, Washington, DC, USA, 1978.

23. R. P. C. Morgan, D. D. V. Morgan, and H. J. Finney, "A predictive model for the assessment of soil erosion risk," Journal of Agricultural Engineering Research, vol. 30, pp. 245–253, 1984. ·

24. D. C. Flanagan and M. A. Nearing, "USDA-Water Erosion Prediction Project (WEPP)," Tech. Rep. 10, National Soil Erosion Research Laboratory, USDA-ARS-MWA, West Lafayette, Ind, USA, 1995.

25. J. G. Arnold, R. Srinivasan, R. S. Muttiah, and J. R. Williams, "Large area hydrologic modeling and assessment part I: model development," Journal of the American Water Resources Association, vol. 34, no. 1, pp. 73–89, 1998.

26. R. P. C. Morgan, J. N. Quinton, R. E. Smith et al., "The European Soil Erosion Model (EUROSEM): a dynamic approach for predicting sediment transport from fields and small catchments," Earth Surface Processes and Landforms, vol. 23, no. 6, pp. 527–544, 1998.

27. R. L. Bingner and F. D. Theurer, "AnnAGNPS: estimating sediment yield by particle size for sheet & rill erosion," in Proceedings of Sedimentation: Monitoring, Modeling, and Managing, 7th Federal Interagency Sedimentation Conference, vol. 1, pp. 1–7, Reno, Nev, USA, March 2001.

28. D. M. Fox and R. B. Bryan, "The relationship of soil loss by interrill erosion to slope gradient," Catena, vol. 38, no. 3, pp. 211–222, 2000.

29. P. I. A. Kinnell, "AGNPS-UM: applying the USLE-M within the agricultural non point source pollution model," Environmental Modelling and Software, vol. 15, no. 3, pp. 331–341, 2000. ·

30. K. J. Lim, M. Sagong, B. A. Engel, Z. Tang, J. Choi, and K.-S. Kim,

"GIS-based sediment assessment tool,"Catena, vol. 64, no. 1, pp. 61–80, 2005.

31.   G. B. Tesfahunegn, P. L. G. Vlek, and L. Tamene, "Management strategies for reducing soil degradation through modeling in a GIS environment in northern Ethiopia catchment," Nutrient Cycling in Agroecosystems, vol. 92, no. 3, pp. 255–272, 2012.

32.   L. Tamene, S. J. Park, R. Dikau, and P. L. G. Vlek, "Analysis of factors determining sediment yield variability in the highlands of northern Ethiopia," Geomorphology, vol. 76, no. 1-2, pp. 76–91, 2006.

33.   EMA, Ethiopian Mapping Agency, "Ethiopia 1:50000 topographic maps: Aksum Sheet," Tech. Rep., Ethiopian Mapping Agency, Addis Ababa, Ethiopia, 1997.

34.   FAO, Food and Agriculture Organization of the United Nations, "The Soil and Terrain Database for Northeastern Africa," FAO, Rome, Italy, 1998.

35.   T. M. Dinka, Application of the Morgan, Morgan Finney Model in Adulala Mariyam Watershed, Ethiopia: GIS based erosion risk assessment and testing of alternative land management options [M.S. thesis], Wageningen University, Wageningen, The Netherlands, 2007.

36.   S. J. Birrell, K. A. Sudduth, and N. R. Kitchen, "Nutrient mapping implications of short-range variability," in Proceedings of the 3rd International Conference on Precision Agriculture, P. C. Robert, R. H. Rust, and W. E. Larson, Eds., pp. 206–216, America Society of Agronomy, Crop Science Society of America, Soil Science Society of America, Madison, Wis, USA, 1996.

37.   D. W. Franzen, A. D. Halvorson, J. Krupinsky, V. L. Hofman, and L. J. Cihacek, "Directed sampling using topography as a logical basis," in Proceedings of the 4th International Conference on Precision Agriculture, P. C. Robert, R. H. Rust, and W. E. Larson, Eds., pp. 1559–1568, America Society of Agronomy, Crop Science Society of America, Soil Science Society of America, Madison, Wis, USA, 1998.

38.   G. B. Tesfahunegn, L. Tamene, and P. L. G. Vlek, "A participatory soil quality assessment in Northern Ethiopia›s Mai-Negus catchment," Catena, vol. 86, no. 1, pp. 1–13, 2011.

39.   G. W. Gee and J. W. Bauder, "Particle-size analysis," in Methods of Soil Analysis, Part 1, A. Klute, Ed., pp. 383–411, America Society of Agronomy, Soil Science Society of America, Madison, Wis, USA, 2nd edition, 1986.

40.   G. R. Blake and K. H. Hartge, "Bulk density," in Methods of Soil

Analysis, Part 1, A. Klute, Ed., vol. 9 ofAgronomy Monograph, pp. 363–375, America Society of Agronomy, Madison, Wis, USA, 2nd edition, 1986.

41. T. C. Baruah and H. P. Barthakur, A Text Book of Soil Analysis, Vikas, New Delhi, India, 1999.

42. Utset, T. López, and M. Díaz, "A comparison of soil maps, kriging and a combined method for spatially predicting bulk density and field capacity of ferralsols in the Havana-Matanzas Plain,"Geoderma, vol. 96, no. 3, pp. 199–213, 2000.

43. N. A. C. Cressie, Statistics for Spatial Data, Revised Edition, John Wiley & Sons, New York, NY, USA, 1993.

44. J. Triantafilis, I. O. A. Odeh, and A. B. McBratney, "Five geostatistical models to predict soil salinity from electromagnetic induction data across irrigated cotton," Soil Science Society of America Journal, vol. 65, no. 3, pp. 869–878, 2001.

45. D. Chekol, Modeling of hydrology and soil erosion of upper Awash river basin, Ethiopia [Ph.D. thesis], University of Bonn, Bonn, Germany, 2006.

46. P. J. J. Desmet and G. Govers, "GIS-based simulation of erosion and deposition patterns in an agricultural landscape: a comparison of model results with soil map information," Catena, vol. 25, no. 1–4, pp. 389–401, 1995.

47. H. Mitasova, L. Mitas, W. M. Brown, and D. M. Johnston, GIS Tools for Erosion/Deposition Modelling and Multidimensional Visualization, Geographic Modelling and Systems Laboratory, University of Illinois at Urbana-Champaign, Urbana, Ill, USA, 1997.

48. K. M. Turnage, S. Y. Lee, J. E. Foss, K. H. Kim, and I. L. Larsen, "Comparison of soil erosion and deposition rates using radiocesium, RUSLE, and buried soils in dolines in East Tennessee,"Environmental Geology, vol. 29, no. 1-2, pp. 1–10, 1997. ·

49. N. C. Brady and R. R. Weil, Eds., The Nature and Properties of Soils, Prentice Hall, Upper Saddle River, NJ, USA, 13th edition, 2002.

50. J. W. Doran, "Soil health and global sustainability: translating science into practice," Agriculture, Ecosystems & Environment, vol. 88, no. 2, pp. 119–127, 2002. ·

51. H. A. Ahmed, E. S. Gerald, and H. H. Richard, "Soil bulk density and water infiltration as affected by grazing systems," Journal of Range Management, vol. 40, no. 4, pp. 307–309, 1987.

52. P. Behera, K. H. V. Durga Rao, and K. K. Das, "Soil erosion modeling

using MMF model—a remote sensing and GIS perspective," Journal of Indian Society of Remote Sensing, vol. 33, no. 1, pp. 165–176, 2005.

53. O. T. Ande, Y. Alaga, and G. A. Oluwatosin, "Soil erosion prediction using MMF model on highly dissected hilly terrain of Ekiti environs in southwestern Nigeria," International Journal of Physical Sciences, vol. 4, no. 2, pp. 53–57, 2009.

54. V. Garg and V. Jothiprakash, "Sediment yield assessment of a large basin using PSIAC approach in GIS environment," Water Resources Management, vol. 26, no. 3, pp. 799–840, 2012.

55. Q. B. Le, R. Seidl, and R. W. Scholz, "Feedback loops and types of adaptation in the modelling of land-use decisions in an agent-based simulation," Environmental Modelling and Software, vol. 27-28, pp. 83–96, 2012.

56. L. B. Asmamaw, A. A. Mohammed, and T. D. Lulseged, "Land use/ cover dynamics and their effects in the Gerado catchment, Northeastern Ethiopia," International Journal of Environmental Studies, vol. 68, no. 6, pp. 883–900, 2011.

57. P. Kaini, K. Artita, and J. W. Nicklow, "Optimizing structural best management practices using SWAT and genetic algorithm to improve water quality goals," Water Resources Management, vol. 26, no. 7, pp. 1827–1845, 2012.

58. L. Sun, W. Lu, O. Yang, J. D. Martín, and D. Li, "Ecological compensation estimation of soil and water conservation based on cost-benefit analysis," Water Resources Management, vol. 27, no. 8, pp. 2709–2727, 2013.

59. G. B. Tesfahunegn, P. L. G. Vlek, and L. Tamene, "Application of SWAT model to assess erosion hotspot for sub-catchment management at Mai-Negus catchment in northern Ethiopia," East African Journal of Science and Technology, vol. 2, no. 2, pp. 97–123, 2013.

60. Munodawafa, "The effect of rainfall characteristics and tillage on sheet erosion and maize grain yield in semiarid conditions and granitic sandy soils of Zimbabwe," Applied and Environmental Soil Science, vol. 2012, Article ID 243815, 8 pages, 2012.

61. H. Hurni, "Soil formation rates in Ethiopia," Working Paper 2, Ethiopian Highlands Reclamation Studies, Addis Ababa, Ethiopia, 1983.

62. M. J. Machado, A. Perez-Gonzalez, and G. Benito, "Assessment of soil erosion using a predictive model," in Rehabilitation of DegradIng and Degraded Areas of Tigray, Northern Ethiopia, E. Feoli, Ed., pp. 237–248, Department of Biology, University of Trieste, Trieste, Italy, 1996.

Chapter 12

GEO-ENVIRONMENTAL SITE INVESTIGATION FOR MUNICIPAL SOLID WASTE DISPOSAL SITES

Giulliana Mondelli, Heraldo Luiz Giacheti and Vagner Roberto Elis

Institute for Technological Research of São Paulo State, São Paulo State University, University of São Paulo, São Paulo, Brazil

INTRODUCTION

Deactivated dump sites and inadequate sanitary landfills can be a serious potential source of contamination. Due to the large number of these sites, contaminants can be generated and migrate into the ground. For this reason, site investigation programs which consider different types of soils and contaminants are needed. In many countries, sanitary landfills have been recently built according to the engineering standards with satisfactory procedures and following the requirements imposed by the local environmental agencies, including monitoring and proper operations. There is a great concern with the operation of sanitary landfills and with the future of deactivated dump sites located in small and medium-size cities, since many of them were inappropriately constructed and operated. Consequently, the main surficial streams and groundwater aquifers, subsoil and air become vulnerable to contamination and pollution. A continuous site investigation program, including in-situ and laboratory tests are necessary to identify the typical soil profile, hydrogeological characteristics and background chemical values. The site surroundings and the subsoil contamination plume can be characterized based on geotechnical, geochemical and mineralogical techniques. The demand for the geo-environmental site investigation of contaminated and noncontaminated sites intended for future sanitary landfill planning has substantially increased in the last few years due to the lack of space in metropolitan areas. Environmental agencies from several countries have proposed different site investigation methodologies in order to diagnose and confirm different contamination levels in sites with diverse physical characteristics in order to guide the remediation plan whenever it is necessary. The experience already achieved on site investigations has shown that the best methodology is site specific, and depends on subsoil and chemical contaminants; geotechnical, geological and hydrogeological aspects; evolution of the contamination plume and the possible risks it poses. Several field and laboratory investigation techniques (direct and indirect) have been proposed and used. Sometimes one technique is more suitable than another depending on the physical and natural characteristics of the site. Countries with recent concerns over the environment tend to adapt the experience gained from the more developed countries with the reality of their own environmental laws, economy, industrialization, size, etc. Having outlined this scenario, this chapter aims to present and discuss the different tests and steps of a geo-environmental site investigation program proposed for municipal solid waste disposal sites.

GEO-ENVIRONMENTAL SITE INVESTIGATION STEPS

The various steps that can be included during a geo-environmental site investigation are discussed in this item and can be systematized in the chart shown in Figure 1, which was prepared considering experience gained on several geo-environmental site characterization campaigns at municipal solid waste disposal sites in Brazil. The major focus of these studies was to assess contamination of medium size municipal solid waste disposal sites, installed over typical Brazilian tropical soils (residual soils and sandstones).

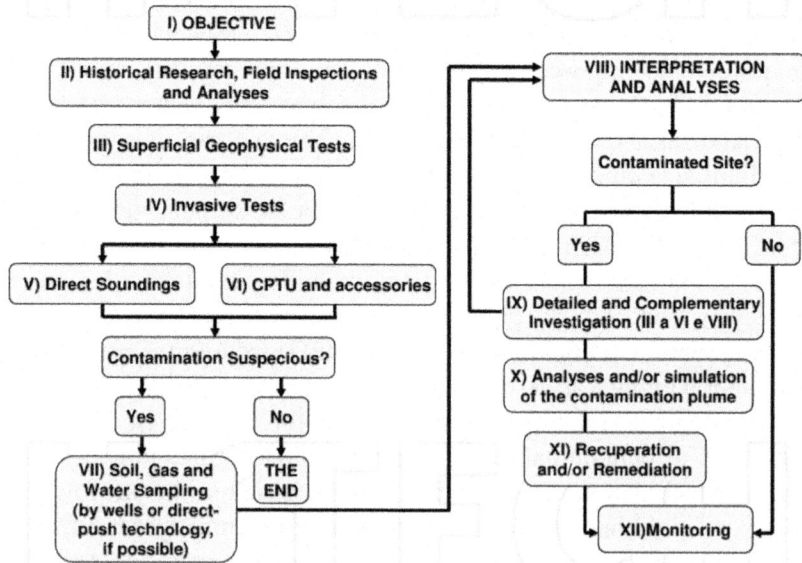

Figure. 1: Proposed geo-environmental site investigation steps for municipal solid waste disposal sites (Mondelli, 2008).

It is important to emphasize that the very first step in any site investigation program is to define the problem. For example, if the objective of the site investigation is just to characterize the local subsoil before the installation of a sanitary landfill, it will probably be necessary to carried out the site investigation until the fifth or sixth step presented in the Figure 1. Depending of the enterprise size or of the natural environmental and anthropological site conditions, the contamination will be assessed in a later step. Therefore, before starting any site investigation test it is fundamental to clearly define the objective of the site investigation program and determine the possible sources and/or targets of contamination. If the sources and/or targets do not exist, there is no sense in continuing the site investigation after the seventh step (Figure 1).

PRELIMINARY INVESTIGATION

Environmental agencies usually recommend gathering historical information and carrying out a site inspection during the preliminary investigation. Compilation of physical and historical data and all the existing information about the site need to be collected before starting any site investigation. This phase (second step in Figure 1) has proved to be very important for the future steps, based on the interpretation of the previous and current topography, hydrology, aerial photographs and historical documents. A good job in this phase can avoid unnecessary tests in the next steps of the site investigation, reducing costs without compromising the quality of the final results. Before installing a new sanitary landfill, closing an old dump or in a regular landfill, or just to investigate the extension of a contamination plume caused by leachate from an irregular municipal solid waste disposal site, it is necessary to contact the responsible company and environmental agencies involved, visit the site and request the following data:

- Location: geographical coordinates, original and current topography, distance from the urban zone, access, surroundings, rivers and airport proximity. Depending on the local regulations, some of these aspects may constrain the construction or extension of new sanitary landfills and may also help the site investigation plan and the final design.

- Municipal waste management: information such as amounts of municipal, hospital, demolition/construction and sweeping/pruning waste generated per person and the locations and manner in which they are disposed of, also needs to be checked. Furthermore, some questions need to be addressed, like: "How long the current landfill will continue to operate? Has it obtained operating and/or commissioning licenses? Is there a plan for hazardous waste management? What percentage of the waste is recycled by the municipality? Are there any environmental education programs for the local community?"

- Aerial photography: a very complete study of the site to be investigated, its uses and occupational activities and physical surrounds can be done by using aerial photography taken over a period of time up to present day photos.

- Old reports: reports of past site investigations and geological surveys need to be available for consultation and search of the physical media data to be also used or supplemented during the new site investigation phase.

- Weather Aspects: climate data, such as mean annual temperatures, rainfall, radiation, etc for the city or that region where the site investigation will be conducted, should be collected for analysis and calculation of the water balance. The water balance can be carried out using the Thornthwaite and Mather (1955) methodology.

After the analyses and interpretation of all integrated existing data from the site to be investigated, a Preliminary Investigation Report can be submitted. Therefore, the site investigation plan will be planned and conducted based on all information collected on site, facilitating the achievement of the final objective of the site investigation, in order to reduce time and costs. Sometimes, this next phase is called Confirmatory Investigation, when suspicion of contamination occurs. More recently, this phase can integrate the use of different field and laboratory tests. The most well known and widely used tests are surficial geophysics (non-invasive and indirect techniques), followed by piezocone tests (CPTU) (invasive and indirect/direct techniques) with special sensors and samplers for environmental purposes, and groundwater level monitoring wells. Accordingly with Figure 1, some of these techniques, which are more frequently used for geo-environmental site investigation purposes, will be briefly discussed.

GEOPHYSICAL SURVEY

Geophysical methods, particularly electrical techniques, can be used to study those different environmental characteristics which are important during the site characterization for waste disposal and the monitoring of migratory contamination plumes. The major advantage of these methods is that the measurements of soil and contaminant properties are indirect, but it can also be a disadvantage too. It allows a quick investigation of a large area avoiding direct contact with contaminants. Direct techniques based on previous tests, such as simple reconnaissance boring with Standard Penetration Test (SPT) and monitoring through groundwater wells and piezometers, can be used and are considered a traditional practice in developing countries. In this context, much effort has been concentrated on measuring electrical resistivity, a property that is highly dependent on lithology and contamination, which are essential factors to be detected during a geo-environmental site investigation. The major interest in measuring electrical resistivity in a geo-environmental site investigation of contaminated sites is due to the fact that it is an indirect measurement that can be achieved by means of subsurface and/or surface tests, which is also sensitive to temporal variations of the physical environment. This also allows it to be used as a monitoring technique and not only as a method for the investigation and initial reconnaissance of the subsoil. The following geological, geotechnical, hydrogeological and environmental characteristics can be assessed using geophysical methods: (a) rock depth; (b) discontinuities; (c) changes of soil texture; (d) groundwater level; (e) groundwater flow; (f) presence and threedimensional distribution of waste; (g) contaminated soil; (h) contaminated groundwater and plume shape. The first five characteristics are essential for the assessment of any waste disposal sites. All these submitted characteristics are recommended in the geoenvironmental monitoring of a waste disposal facility. The geophysical methods which present good results for these applications are the Resistivity and the Low Frequency Electromagnetics (Ground Conductivity Meters).

Resistivity

The Resistivity Method applies an artificial electrical current I, introducing two electrodes into the ground (A and B), with the objective of measuring the differential potential ΔV generated by two other electrodes (M and N) in the electrical current flow extremities (Figure 2). This arrangement allows the calculation of apparent resistivity ρa in subsurface, using the following equation:

$$\rho_a = K.\frac{\Delta V}{I} \text{ (ohm.m)}$$

(1)

where K is a geometrical factor that depends on the electrodes position in the ground and can be calculated for any electrode arrangement using the following general expression:

$$K = \frac{2\pi}{(1/AM)-(1/AN)-(1/BM)+(1/BN)} \text{ (m)}$$

(2)

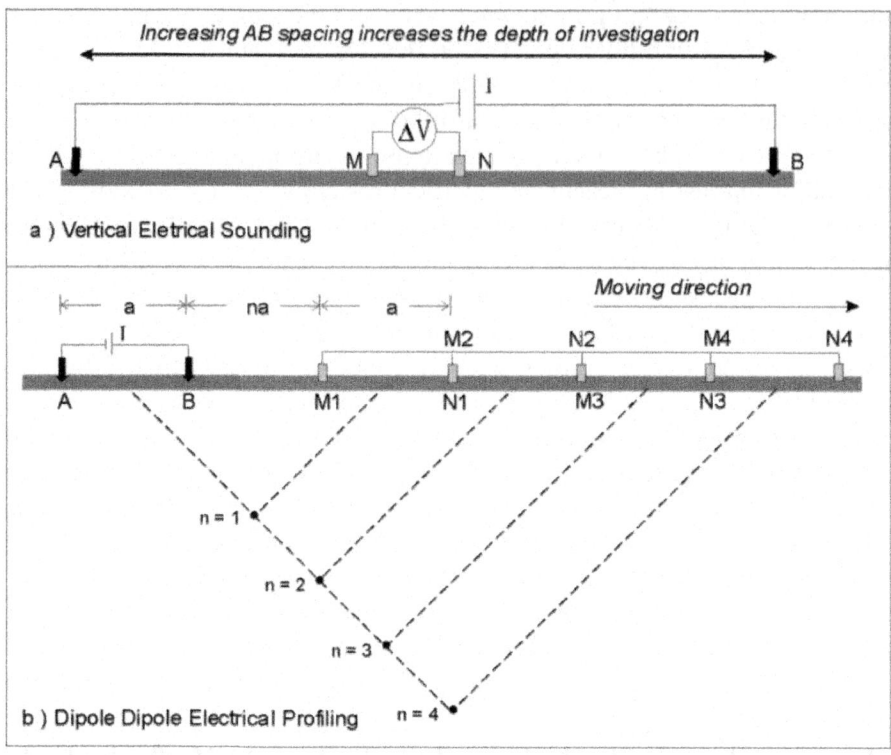

Figure. 2: Configuration of 1D and 2D resistivity surveys: (a) Vertical Electrical Sounding and (b) Dipole Dipole Electrical Profiling.

The soil and rock resistivity values are affected mainly by four factors: mineralogy composition, porosity, moisture, amount and nature of dissolved salts. The most important factors are the porous water content and the salinity content in the porous media. The increase of these two factors tends to decrease the resistivity values. These characteristics allow a good applicability of this geophysical method of application in environmental and hydrogeological studies, where inorganic contaminants occur. In the specific case of municipal solid waste disposal sites, the formation of leachate with elevated concentration of ions, characterizing the polluted areas with very low resistivity values.

The equipment used for the resistivity method consists basically of a controlled electrical current source emission and differential potential measurements. The source potential can range from some kilowatts up to hundreds of watts. This equipment can work using direct or alternating current supply with low frequency, preferably lower than 60 Hz (Telford et al., 1990). The Resistivity Method encompasses various techniques for the application of field tests, which basically consist of vertical electrical sounding (1D resistivity survey) and electrical profiling (2D resistivity imaging) and includes a wide variety of possible electrode configurations (Schlumberger, Wenner, Dipole Dipole, Pole Dipole, Lee and other), rendering the method highly versatile. Currently, 3D tests are possible with the use of multielectrode equipment or a series of 2D profiles grouped and equidistant (Loke & Barker, 1996; Dahlin et al., 2002). Figure 2 presents electrical soundings and dipole dipole electrical profiling field arrangements. As the subsoil is not homogeneous, for any measurement point an apparent resistivity value is obtained. Appropriate software is used for data acquisition and interpretation, for obtaining the electrical stratigraphy of the subsoil (1D survey) or the resistivity cross-section (2D survey). There are several possibilities for carrying out these tests allowing the use of different techniques and arrangements, depending on the site investigation objectives. Usually the most important objective of a geo-environmental characterization is to delineate contamination plumes. The vertical electrical soundings can give important information on site characterization of waste disposal sites. The distribution of different geomaterials on the subsurface profile (several soil layers, waste and rock substrate top) and the depth of the saturated zone can be estimated using vertical electrical soundings. Figure 3 presents the results and the interpretation of a vertical electrical sounding carried out to expand the knowledge of the stratigraphy profile and for defining the position of the saturated zone. The combined results obtained from several vertical electrical soundings allow the construction of a contour map of the groundwater level showing the direction of the groundwater flow. Waste disposal sites can be mapped using several vertical electrical soundings (VES) points or using electrical profiling (continuous way). The VES are more appropriate to vertical subsoil profile definition (waste thickness and landfill bottom). The electrical profiling gives a better delineation of the waste disposal dimensions. 2D inversion surveys of electrical profiling show both lateral and vertical variations of the materials in the cross-section. Figure 4 presents two municipal solid waste trenches excavated on sandy geological substrate delineated by electrical Resistivity Method results (resistivity values lower than 20 ohm.m characterize the places filled with municipal solid waste). Figure 5 shows the results of a survey performed in a contaminated site downstream from a landfill in Brazil. This site was investigated using a 3-D resistivity imaging technique. The purpose of this investigation was to detect and delineate the contamination plume produced by the waste and to acquire detailed information about the affected area. The data set consisted of a series of parallel electrical profile data acquired with the dipole-dipole arrangement, and was inverted as a complete 3-D survey. The resistivity model identifies the disposed waste and the contaminant, marked by the isosurface values lower than 20 ohm.m. The results indicate the presence of a contamination plume and its preferred path. Monitoring wells were installed in the affected area and their chemical analysis confirmed the influence of contaminants.

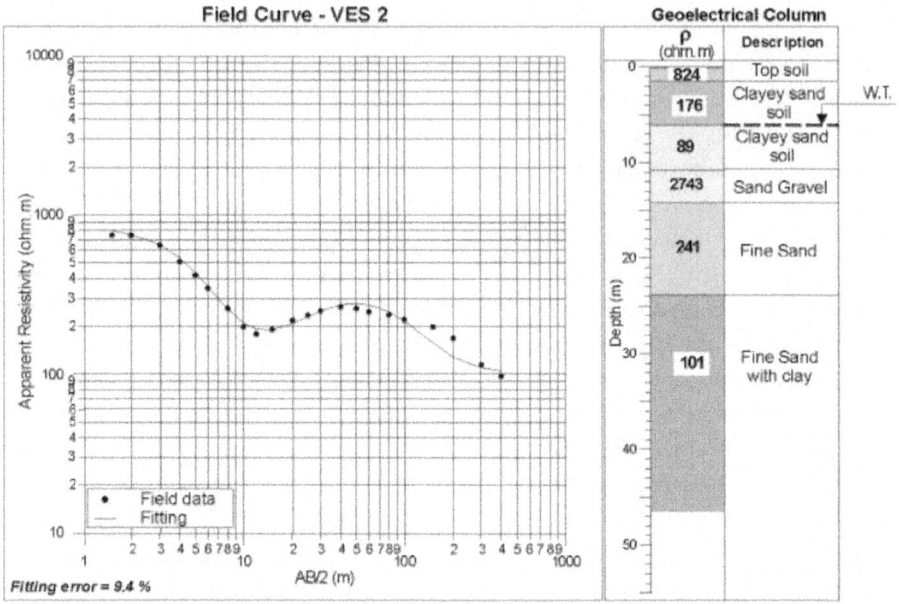

Figure. 3: Results of VES for studying the stratigraphic profile and position of the saturated zone (W.T.).

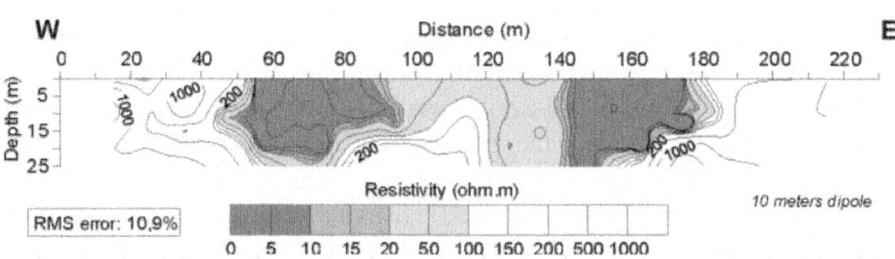

Figure. 4: Resistivity cross-section of waste disposal area. Lower resistivity zones characterize two trenches filled with municipal solid waste.

Low Frequency Electromagnetics (Ground Conductivity Meters)

The Electromagnetic Methods (EM) involves low frequency electromagnetic field propagation. When an alternating current (AC) is established in a wire placed on the ground surface, electrical currents flow in subsurface conductors. This process is known as electromagnetic induction and constitutes the basis of the operation of low frequency electromagnetic methods.

This method is very fast and it uses simple equipment easily operated. It explains the extensive application of this method for geo-environmental studies. The equipment used can be generally called ground conductivity meters. It consists of two coils (transmitter and receiver). The transmitter coil emits a primary magnetic field H_p, which induces electrical currents on the subsurface, generating a secondary magnetic field H_s (Figure 6). The combination of these two fields is measured by the receiver coil. Under certain conditions it is known that there is a linear relationship between the modules of the two fields (McNeil, 1980), technically defined as "low induction number operation". Accordingly, the apparent conductivity can be calculated by σ_a:

$$\sigma_a = \frac{4}{\varpi\mu_0 s^2}\left(\frac{H_s}{H_p}\right) \ (mS/m)$$

(3)

where ω = angular frequency, $2\pi f$; μ_0 = vacuum magnetic permeability; s = intercoil spacing.

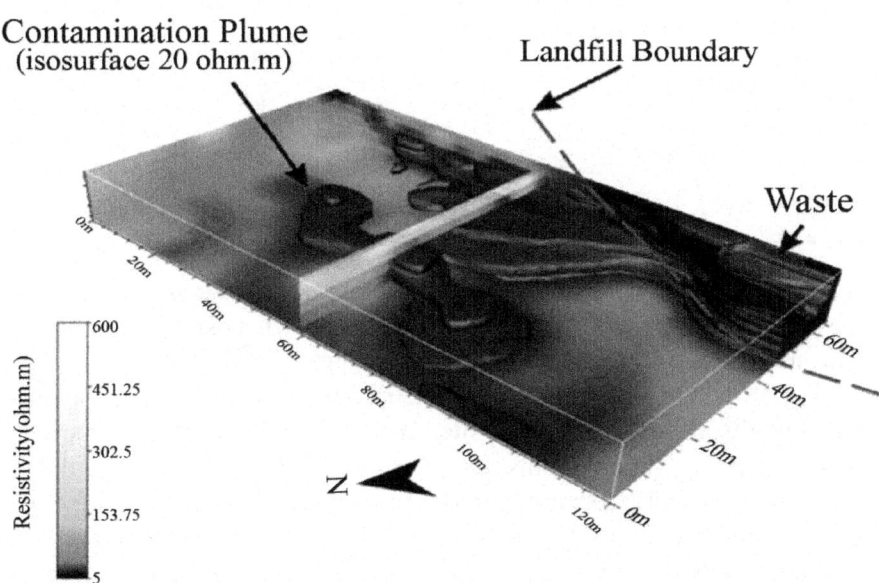

Figure. 5: 3D resistivity model of a contaminant plume generated by a municipal solid waste disposal site. The isosurface of 20 ohm.m delineates the contaminant plume, which is moving in a northwest direction.

The equipment is built to allow a direct reading of the apparent conductivity. The most commonly used equipment at present time was designed to explore pre-defined depths, between 7.5 to 60.0 meters, depending on the coils orientation (vertical loop and horizontal loop modes), the operating frequency and the intercoil spacing. The field tests are usually conducted in profiles, which, due to convenience of operation and transportation of the equipment, are carried out very quickly. The apparent conductivity data can be plotted on several profiles and depending on the distance, a set of profiles can form a site map. The interpretation of these data is basically qualitative, but nowadays there is inversion software designed for quantitative interpretation of conductivity profiling data (Monteiro Santos, 2004).

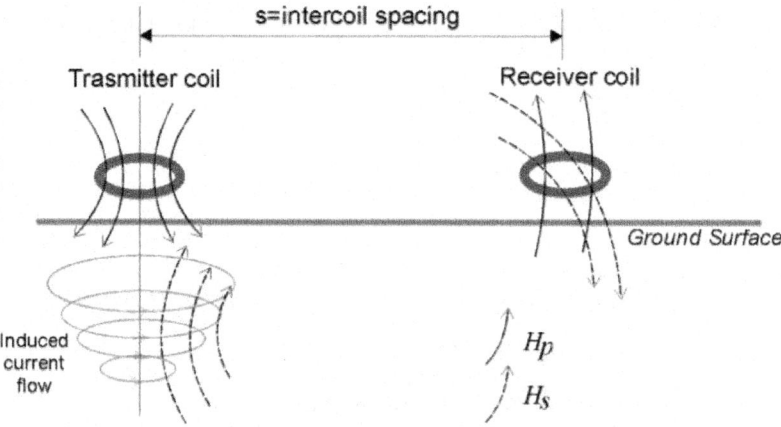

Figure. 6: Operating principle of the Ground Conductivity Meter.

The method is a cost effective and low time-consuming tool for investigating contamination generated by waste disposal sites, and allows delineation of the area affected by contaminants. Figure 7 shows the mapped out area of the contamination plume delineation, which has been created by a municipal solid waste disposal site. The MSW boundaries are marked in a broken white line (black points are measurement stations). The northwestern part of the landfill presents low electrical conductivity, characterizing a granitic base. The site is occupied by waste deposited onto tertiary stream sediments, and the groundwater level appears at around 5 meters depth. The contours of the waste disposed at the site can be identified based on the values of apparent electrical conductivity higher than 50 mS/m. The contamination plume can also be observed in this figure, which flows to the West, marked by high values of apparent conductivity outside of the waste filled area.

INVASIVE TESTS

In some particular sites it is difficult to identify the subsoil properties contrasts based only on geophysical data or index properties, especially in a very heterogeneous municipal solid waste (MSW) disposal site on unsaturated tropical soils. The proper site investigation in these sites relies on invasive tests, including soil, groundwater and direct gas sampling. The modern approach for geo-environmental site characterization using invasive tests relies on the piezocone technology. The piezocone and it accessories are used as a screening tool for stratigraphic logging of geotechnical and chemical measurements. This approach often allows identifying potentially critical zones which may require more specific tests to measure or monitor contaminants based on specific sampling and/or monitoring.

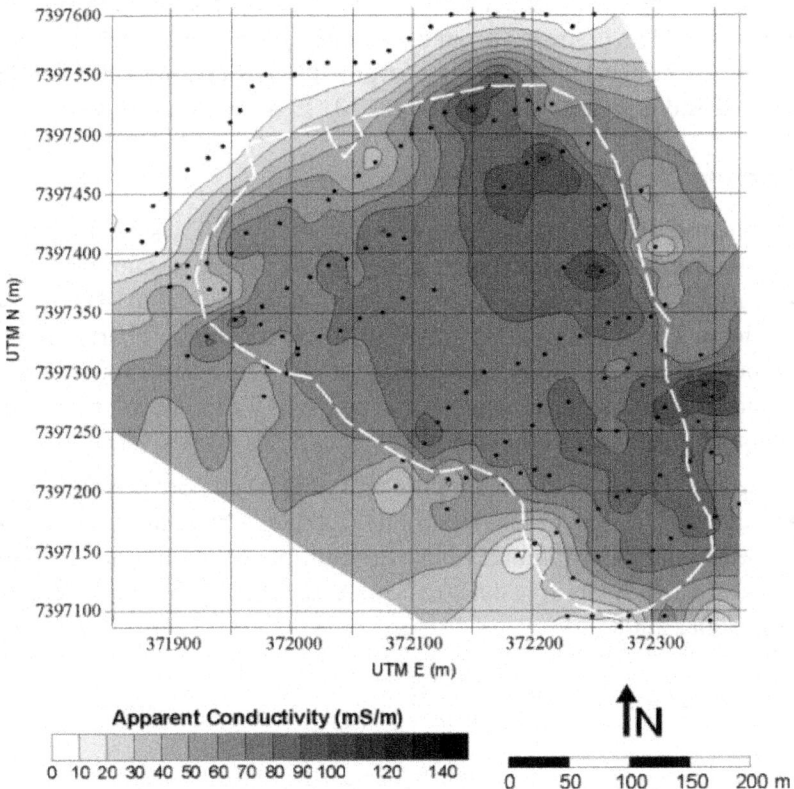

Figure. 7: Apparent conductivity map of a waste disposal site. Theoretical depth of exploration is 7.5 meters. The broken white line marks the limits of the deposit. Note the good match of the boundaries of the deposit with the high apparent conductivity values (> 50 mS/m).

Piezocone Test and Accessories

The Piezocone

Piezocone penetration test (CPTU) is a standard site investigation tool currently used for geotechnical site characterization (Lunne et al, 1997). According to Shinn II & Bratton (1995), the heart of a modern site characterization approach uses the piezocone penetration test. The CPTU test uses a standard instrument probe (ASTM D 5778-07) with a 60° apex and typically 35.68 mm diameter (10 cm2 area) fitted on the end of a series of rods (Figure 8). The probe is pushed into the ground at an approximate constant rate of 2 cm/sec by a hydraulic pushing source, such as a standard drill rig or cone pushing vehicle. As the cone is advanced, the forces measured by the tip and friction sleeve will vary with the material properties of the soil being penetrated. Tip resistance (qc), friction sleeve stress (fs) and dynamic pore-pressure (U) response are measured by calibrated electrical instruments. All channels are continuously monitored and typically digitized at 10, 20, 25 or 50 mm intervals.

Piezocone Data and Interpretation

Piezocone test data is gathered by a computer which allows the user to carry out straightforward post-investigative analysis. The three parameters (q_c, R_f and U) in various combinations, such as friction ratio ($R_f = (f_s/q_c)100\%$), are used to delineate site stratigraphy by using soil classification charts, as proposed by Robertson et al. (1986) and presented in Figure 9. Empirical and semi-theoretical correlations are available in relevant literature in order to estimate mechanical properties of the soil.

Measurement of equilibrium pore pressure at full pore pressure dissipation allows quantifying the vertical hydraulic gradients by using a single sounding and groundwater flow regime, whenever multiple soundings are available. The pore pressure dissipation test also allows estimation of hydraulic conductivity (Campanella, 2008).

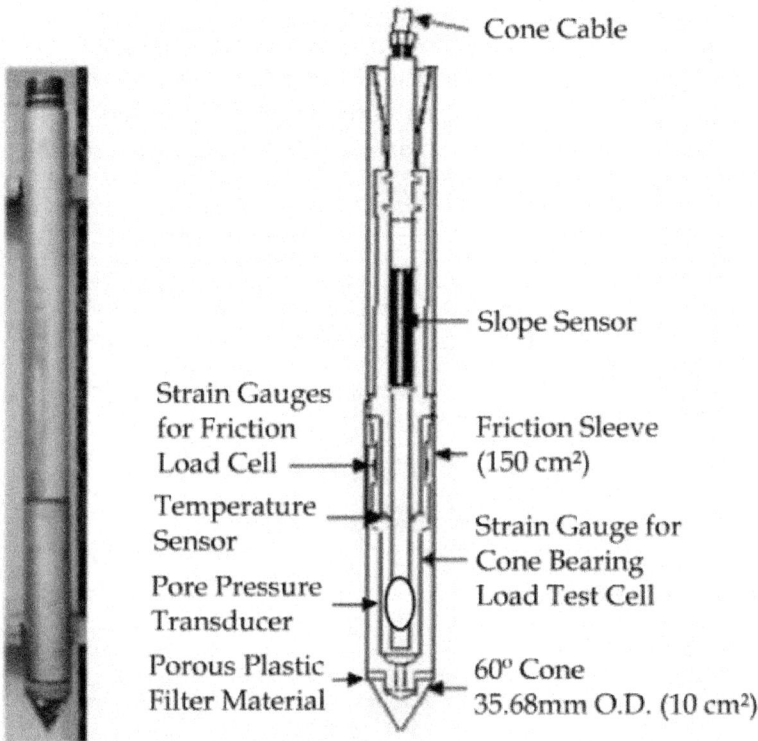

Figure. 8: Schematic drawing and photo of a piezocone probe (modified from Davies & Campanella, 1995).

Accessories

As pointed out by US EPA (1989), a geo-environmental site characterization requires information on the chemical distribution and sources (s)/receptor(s) for potential or existing contaminants. The piezocone technology for geo-environmental application included specific sensors for temperature, resistivity, pH, laser-induced fluorescence, among others. This technology also includes samplers to be used together with the piezocone for sampling soils, water and gas (Lunne et al, 1997). Robertson & Cabal (2008) present in details the main piezocone accessories available for geo-environmental site characterization. The use of the piezocone test and accessories for geo-environmental applications essentially creates no cuttings, produces little disturbance, and reduces contact between field personnel and the contaminants, as the penetrometer push rods can be decontaminated during retrieval (Robertson, 1998). The major limitation of the piezocone test is the impossibility of carrying out the test in gravel.

Resistivity Piezocone

Resistivity is one of the piezocone accessories which is called the resistivity piezocone (RCPTU) test. It measures the electrical resistance of a current flow in the ground. This additional information is extremely useful due to the significant effects that dissolved and free product constituents have on bulk soil resistivity (Campanella, 2008).

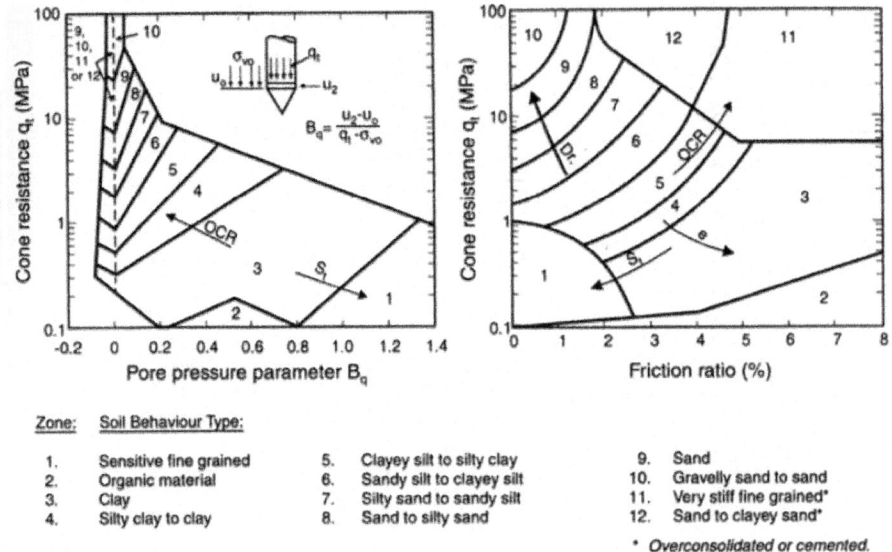

Zone:	Soil Behaviour Type:				
1.	Sensitive fine grained	5.	Clayey silt to silty clay	9.	Sand
2.	Organic material	6.	Sandy silt to clayey silt	10.	Gravelly sand to sand
3.	Clay	7.	Silty sand to sandy silt	11.	Very stiff fine grained*
4.	Silty clay to clay	8.	Sand to silty sand	12.	Sand to clayey sand*

Overconsolidated or cemented.

Figure. 9: Piezocone soil behavior type charts proposed by Robertson et al (1986).

The preparations involved in the RCPTU test are similar to those of any other CPT test (Robertson & Campanella, 1988). The only additional procedure is the connection of a signal generator to the data acquisition system for controlling the current level and frequency of the electrical resistance measurements. Weemees (1990) discusses the importance of working with alternating current and sufficiently high frequencies to prevent polarization of the external electrodes. A frequency of 1000 Hz is usually used. Figure 10 presents a Wenner-type resistivity piezocone (RCPTU) with an array of four electrodes. Resistivity measurements are taken with the inner electrodes and the current is applied through outer electrodes. These measurements are digitally recorded at 25 mm intervals, providing essentially continuous in-situ data sampling in addition to all the other standard CPTU measurements. Measurements of bulk resistivity trends indicate whether some dissolved or free product exists below or above the background values. The background values are usually established from RCPTU tests carried out on-site. According to Campanella (2008), the areas where readings are very different (anomalies) from the background values are then further evaluated with appropriate groundwater sampling at discrete depths for detailed chemical analysis. Considerable practical value is gained from the fact that the measured resistivity in saturated soil is almost totally governed by the pore fluid chemistry.

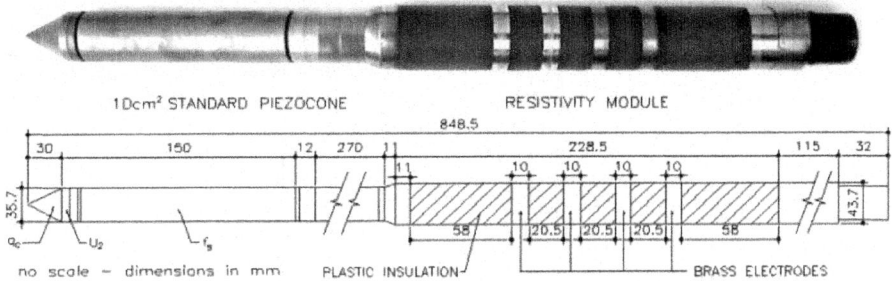

Figure. 10: Schematic representation and photo of a resistivity piezocone probe with a fourelectrode arrangement (Mondelli et al., 2007).

Samplers

A large number of samplers for gas, water and soil sampling have been developed for pushing into the ground using the same equipment that is used for the piezocone test. These samplers are designed for either one-time samples or as monitoring wells. Robertson & Cabal (2008) describe in details some samplers available for geo-environmental site investigation and they will be briefly presented in the follow items.

Soil Sampling

A wide variety of push-in discrete depth soil samplers are available and most of them are based on designs similar to the Gouda soil samplers. Robertson & Cabal (2008) describe some soil samplers including the Gouda type sampler. It is pushed into the ground at the desired depth in a closed position and it has an inner cone tip that is retracted to the locked position leaving a hollow sampler with small diameter (25 mm) stainless steel or brass sample tubes. The hollow sampler is then pushed to collect a sample. The filled sampler and push rods are then retrieved to the ground level. Figure 11 shows a schematic drawing of a typical CPT based soil sampler (Gouda type), as described in detail by Robertson & Cabal (2008).

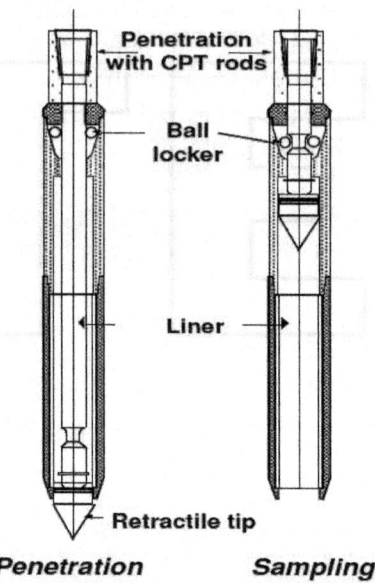

Figure. 11: Schematic drawing of a typical (Gouda type) CPT based soil sampler.

Water Sampling

Robertson & Cabal (2008) affirm that the most common direct-push, discrete depth, in-situ water sampler is the Hydropunch and to a lesser extent, the BAT, Simulprobe and Waterloo Profiler. The Hydropunch sampler and its variations is a simple sampling tool that is pushed down to the desired depth and the push rods withdrawn to expose the filter screen and is described in detail by Robertson & Cabal (2008). A modification of the commercially available BAT groundwater sampler is recommended by Campanella (2008) for obtaining in-situ pore fluid samples and it is presented in Figure 12. This author discribes the original BAT system, which consists of a sampling tip that is accessed through sterile evacuated glass sample tubes and a double-ended hypodermic needle set-up pushed through septum seals. The tube sampler is lowered either by cable or electrical wire depending upon whether a pore fluid sample is taken with or without a pressure test being carried out. Acording to Campanella (2008), the BAT probe is also able to take pore gas samples for collecting volatile contaminants.

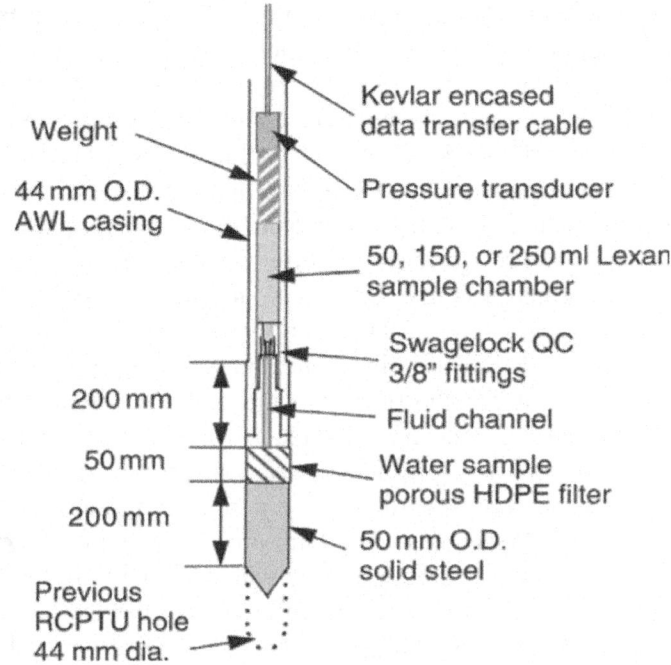

Figure. 12: UBC modified BAT groundwater sampler (Campanella, 2008).

Gas Sampling

Gas samples can be collected in a similar way to the one previously described for groundwater samples. The most common gas sampler that uses the direct-push, discrete depth, in-situ method is the hydropunch type sampler. Robertson & Cabal (2008) also describes this sampler in detail, which is pushed to the required depth, thus the filter element is exposed and a vacuum is applied to draw a vapor sample to the surface. Special disposable plastic tubing is used to draw the sample to the surface.

Example of Using RCPTU Test and Samplers in a MSW Disposal Site

An example of using the piezocone technology to assess contamination of a MSW disposal site from Brazil, by using different techniques is going to be briefly presented based on Mondelli's (2008) research. The site can be considered to be a controlled dump site because it is a planned landfill that incorporates some of the features of a sanitary landfill. According to Mondelli et al. (2007), the site's geology has sandstone from the Adamantina and Marilia Formations, covered by alluvial sandy soils or colluvial clayey sands. Residual soils from sandstone are found underneath these layers. The hydraulic conductivity of the soil was found to vary from 10-7 to 10-6 m/s. The depth of the groundwater level is about 5 meters below the base of the landfill. In order to protect the subsurface and the shallow groundwater table from leakages from the landfill, four 20-cm-thick layers were compacted at the bottom of the landfill using local soil. The upper surface of these layers was coated twice with diluted asphalt emulsion to seal and protect the bottom of the landfill. This procedure was used because it was inexpensive and acceptable for controlled dumps in Brazil when this landfill was established. This site was first investigated using a 3-D resistivity imaging technique, as described by Ustra et al. (2011). Several piezocone and resistivity piezocone tests were carried out on this site. In the first campaign, a four standard piezocone was carried out, measuring cone tip resistance (q_c), sleeve friction (f_s) and pore pressure (u_2). A resistivity piezocone was used in the other campaigns (16 tests). Pore pressure was recorded using a slot-filter filled with automotive grease in all the tests, as suggested by Larsson (1995). A multifunction reaction system equipped with a hydraulic device with a capacity of 200 kN was used to carry out the tests. Details of all the different tests carried out in this site are presented by Mondelli et al. (2007). Two piezocone tests were conducted at the highest parts of the landfill, to obtain a reference profile of the uncontaminated soil. However, resistance to cone penetration was greater than the capacity of the penetration system needed for reaching the groundwater level in this region, where residual soils and sandstone are shallower. Most of the RCPTU tests were performed downstream of the landfill, where sedimentary soils or more homogeneous tropical soils occur. All the RCPTU tests that reached groundwater presented an abrupt reduction in resistivity. This information was useful to help identifying groundwater level (GWL). Two of the RCPTU test results are presented in Figure 13 and will be briefly discussed. The resistivity profiles shown in this figure for the saturated zone, indicate that the resistivity measurements are affected by soil texture and mineralogy. Mondelli et al. (2007) point out that the resistivity values were higher in sandy layers than in clayey layers. At that time the influence of soil type on the resistivity values of the RCPTU-14 and RCPTU-15 tests were not so clear, since the resistivity value found in the RCPTU-14 test was around 50 ohm.m, while that of the RCPTU-15 test was around 20 ohm.m for saturated zone. It indicates a migration of the contamination plume through the sandy layer, based on water sampling using a direct-push sampler at depths of 8.0 to 9.0 m, since it presented low electrical resistivity, which is indicative of the presence of contaminants dissolved in the water. Mondelli et al. (2007) conclude that the RCPTU interpretation for this particular site required collection of soil and groundwater. It was also necessary to adapt a system to measure the electrical resistivity of undisturbed soil samples in the laboratory for better understanding the resistivity variation according to the different saturation conditions and lithological characteristics that occur in tropical regions. This system, as well as the test results, is presented by Mondelli et al. (2010a). The sampling depths were selected based on the interpretation of nearby piezocone and other tests. Mondelli et al. (2010b) present in Figure 14 the relation between the resistivity values of the saturated layers and the results of the characterization tests on soil samples collected using a direct-push sampler, in order to allow identifying and visualizing the layers susceptible to contamination. These results indicate a tendency for increasing values of resistivity with decreasing fines content, clay content or with clay mineral activity. However, layers that do not clearly follow this tendency are also shown in Figure 13, presenting low values of resistivity (between 20 and 40 ohm.m) for the sandier soils. These results demonstrate that values of resistivity ranging from 20 to 40 ohm.m for predominantly sandy layers indicate the presence of leachate in the groundwater. This kind of interpretation goes beyond the merely comparative one of the resistivity values obtained from all the tests discussed by Mondelli et al. (2007), which had diagnosed the presence of contaminants only in tests RCPTU 14 and 15 presented in Figure 13. A certain degree of superposition of values may occur in this type of analysis, since the more clayey layers also show low values of resistivity. Since the flow of contaminants tends to occur between the sandier layers (more permeable), the problem can be simplified in an attempt to identify only these layers.

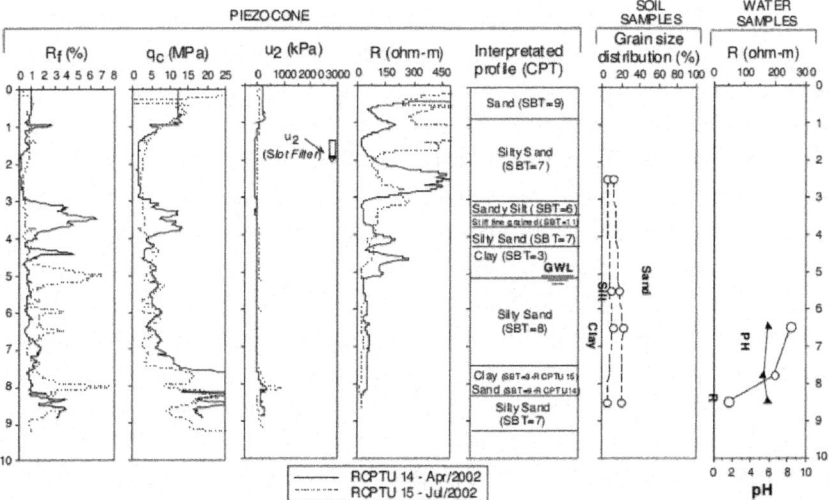

Figure. 13: RCPTU test results, grain size distribution and groundwater analyses for a MSW disposal site, adjacent to the edge of the landfill (Mondelli et al., 2007).

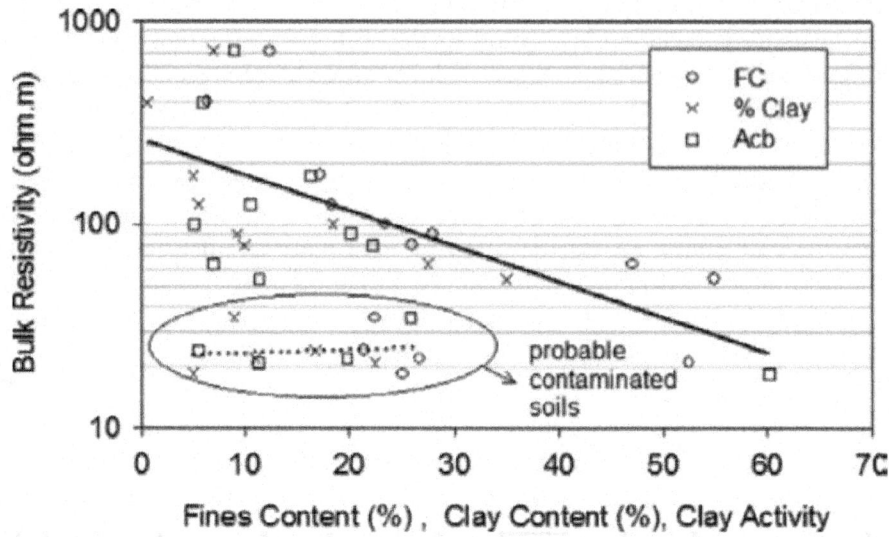

Figure. 14: Electrical resistivity in saturated soil samples with characterization of the fine fraction of soils for a MSW disposal site (Mondelli et al., 2010b).

Mondelli et al. (2010b) studies concluded that the resistivity piezocone test helped identifying contaminated zones. However, this technique presented some limitations for tropical soils, since the groundwater level is sometimes deeper than the penetrable capacity of the cone. The RCPTU tests were useful for detailing stratigraphic soil profile. The relation between corrected point resistance (q_t) and friction ration (R_f) allowed identifying the highly heterogeneous stratigraphic profiles of alluvium and colluvium. This identification for the residual soils was restricted to the behavior of the materials in response to the penetration of the piezocone, requiring soil sampling. The resistivity profiles were useful for the identification of the position of the groundwater level, and were sensitive to variations in the soil's texture and its saturation. A good relationship between the soil behavior index (I_c) and the fines content in the soil samples collected adjacent to the landfill helped the identification of contaminated regions of the saturated zone. The proper interpretation of piezocone tests enabled the identification of those layers more susceptible to contamination, which was confirmed by the chemical analyses of the groundwater samples collected from the monitoring wells. The main conclusion of this study is that the interpretation of RCPTU tests is not as straightforward as it is for sedimentary soils because the tropical soils genesis complexity.

LABORATORY TESTS FOR GEO-ENVIRONMENTAL SITE CHARACTERIZATION

Laboratory tests can improve the interpretation of the in-situ test results, complementing, controlling and detailing the geo-environmental site investigation. A complete geoenvironmental site investigation using current engineering practice is usually difficult to carry out because it is expensive and normally requires authorization and multidisciplinary teams. In addition, an appropriate testing campaign is time consuming and only a few private companies can do it. Limited publications on monitoring and modeling of the contamination plume caused by MSW disposal sites are available. The most commonly known case histories (Cherry et al., 1983; MacFarlane et al., 1983; Mackay et al., 1986; Kjeldsen et al., 1998; Zuquette et al., 2005) have installed more than a hundred monitoring wells, with multilevel soil and groundwater sampling, and are taken from the same sites which have been studied for years. All these studies show that geophysical tests were carried out before the intrusive tests, which were also performed to confirm contamination.

Christensen et al. (2000) explain that, contrary to the specific and known sources of contamination, – such as leaking fuel tanks (hydrocarbons) and chlorinated solvents – the municipal solid waste disposal sites are large, heterogeneous and receive various types of waste over time, have different flow pathways, hydraulic gradients and contamination plumes. Moreover, the size of the landfills will often prohibit removal of the waste and source of the leachate, suggesting that the landfill body and the leachate plume should be considered as a continuum in the context of natural attenuation. In Brazil, most of the laboratory tests for sanitary landfills are preventive, for bottom or cover liner design, using low permeability compacted tropical soils with or without flexible membrane liners (Ritter, 1998; Boscov et al., 2001; Leite & Paraguassú, 2002; Stuermer, 2005; Azevedo et al., 2006). This approach does not consider the older waste deposits, where the contaminants are in direct contact with the natural subsoil. In Brazil, there are some researches on geo-environmental site investigation of industrial and municipal solid wastes disposal sites, using geophysics or boreholes to detect contamination plumes (Grazinolli et al., 1999; Elis & Zuquette, 2002; Lago et al., 2003; Anirban et al., 2004; Bolinelli Jr., 2004; Porsani et al., 2004;). The in-situ techniques permit assessing the natural conditions of the soils, while the laboratory tests provide greater detail, but in minor scale, trying to represent the field conditions based on soil and water/gas samplings. Those difficulties in defining a typical soil profile, hydrogeological parameters and background resistivity (or conductivity) values due to geological complexity (like tropical sites, for example) demand more and varied site investigation techniques. After a better understanding of the site history, topography, hydrogeology and the landfill construction characteristics, laboratory tests can be carried out using disturbed and undisturbed soil samples, in order to investigate a new site for the installation of a sanitary landfill, or to confirm a pre-indicated contamination or to provide support for future remediation actions. The typical soil profile, hydrogeological characteristics and background values can be confirmed and obtained also using laboratory tests. The soil can be characterized using geotechnical, geochemical and mineralogical techniques. Batch sorption and column leaching tests can be carried out using those geomaterials found at the study site. The landfill's leachate can be used to estimate the dispersion and retardation coefficients, for future contamination flow transport modeling. Electrical resistivity values and mineralogical constituents of non-contaminated soils can be determined in the laboratory.

Soil Characterization

The characterization tests are fundamental for the classification of soils, due to their complex geological formation, and even more when performing a geo-environmental site investigation, where the interaction between the contaminants and the physical environment usually occurs. The more usual laboratory characterization tests to investigate a municipal solid waste disposal sites are:

- Geotechnical Properties (Lambe & Whitman, 1969): classical laboratory tests for physical soil characterization are essential for geo-environmental site investigation. Indexes like in-situ moisture content (w), natural unit weight (g), specific gravity (Gs), grain size distribution with and without sodium hexametaphosphate for determination of clay, silt, sand and gravel contents, liquid limit (LL) and plastic limit (PL), shall be determined using disturbed and undisturbed soil samples. Also, permeability / hydraulic conductivity (K) tests are necessary for a good monitoring and pollution prevention of a waste disposal site or for future sanitary landfill, respectively. Depending on the groundwater position and the characteristics of the site, tests to determine the soil retention curve can also be important in a site investigation and monitoring program.

- Organic Matter Content (OMC): OMC is considered an important property of the soil solid phase, as it is responsible for retaining a good part of the organic contaminants and leachate ions, together with clay minerals and hydroxides. OMC can be determined by incineration of the soil sample in a muffle, heating to temperature of 440oC, as described in ABNT NBR 13600 (1996).

- Blue Methyl Adsorption: this test allows the estimative of the cation exchange capacity (CEC), specific soil surface (SE) and activity of clay minerals. This test also indicates the predominant type of clay mineral matrix in fines content of the soil sample. Lan (1977) and Pejon (1992) describe this test method in detail.

- X-Ray Diffraction (XRD), Differential Thermal Analysis (DTA) and Gravimetrical Thermal Analysis (GTA): These tests allow characterization of the soil fines content, by heating up the soil sample to 1000oC, in order to identify and quantify different clay minerals, hydroxides and any solid constituent of the soil. Consequently, the tests can be used to identify the soil mineralogy and its geochemical interaction with the landfill, or for assessing natural attenuation of the contaminants.

- pH: Determination of pH of granular materials in deionized water (usually ratio of 1 : 2.5), by measurement of the effective electrochemical concentration of H+ ions in soil solution or waste using a combined electrode immersed in suspension. The test method is that one recommended by US EPA (1993).

- X-Ray Fluorescence (XRF): XRF spectrometry is used to identify elements in a soil and quantify the amount of those elements (metals) present. An element is identified by its characteristic X-ray emission wavelength (λ) or energy (E). The amount of an element present is quantified by measuring the intensity of its characteristic line. It is known that original rock composition can or cannot present natural high concentrations of some metals. For geo-environmental purposes, these conditions cannot exceed safety, health or potability exigencies for the local population. Therefore, it is noted that the levels of these metals in soils and sediments depend on the composition and proportion of them in their original solid phase. The background values need to be diagnosed before or after the waste disposal. Innov-X Systems Alpha Series (2007) presents handheld EDXRF spectrometers, ideally suited for field and laboratory analysis of soil and waste metals.

As example, Figure 15 presents a typical soil profile, characterized for construction of a future waste disposal site for a very small city in the interior of the São Paulo State, Brazil. IPT (2011) carried out a low cost geo-environmental site investigation campaign with the objective of detecting possible contamination caused by an older controlled dump near to the interest site. The results presented in Figure 15 shows the log of the number of blows from Standard Penetration Test (N_{SPT}) up to 16.5 m depth, when the impenetrable layer was reached. During the SPT tests, deformed soil samples were collected, in order to maintain the natural moisture, and were transported to the laboratory. In the laboratory soil samples were identified based on tactile visual inspection. So, "new"soil samples were composed using the same lythology. The collected soil samples were characterized, determining moisture content (w), organic matter content (OMC), pH, electrical conductivity, grain size distribution and XRF analysis using an Innov-X handheld EDXRF spectrometer. Figure 15 shows that the soil profile is formed by a very loose to loose sand, up to about 9.5 m depth. Between 9.5 and 13.0 m the subsoil has a medium compacity (9 < N_{SPT} 20. Residual soil from sandstone was found below 15.0 m depth. The moisture content was around 7-8 %, which changed just when the groundwater level was reached, at 13.0 m depth, increasing to 17 %.

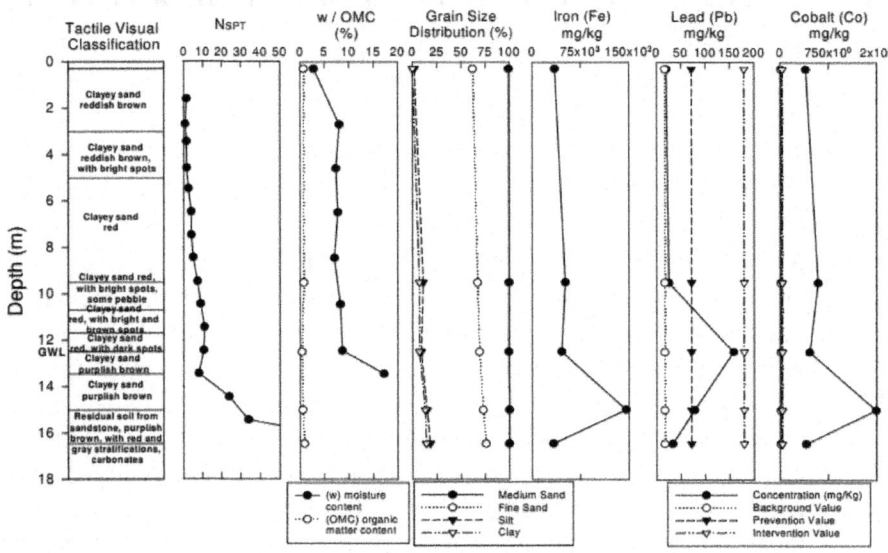

Figure. 15: Soil characterization profiles for a geo-environmental site investigation to install future sanitary trenches in a small city from Brazil.

The local soil is basically constituted of a fine to medium sand. Clay content tends to increase with depth, ranging from 0 to 16%. The organic matter content was found to be very low, ranging between 0.7 and 0.9%, tending to decrease to 0.3 to 0.4%, for the residual soil from sandstone. The geochemical profiles obtained for iron, cobalt and lead are also presented in Figure 15. They have to be compared with the background, prevention and intervention values determined for agricultural use by the Environmental Company from São Paulo State (CETESB, 2005). There is no reference value published for iron in the São Paulo State, Brazil. It is observed that all parameters have reached peak concentrations when residual soil from sandstone is encountered. The cobalt value is greater than the intervention value and lead levels are greater than the reference value throughout all the subsoil profiling. The lead level is also greater than the intervention value for the residual soil. These high levels of Co, Pb and Fe may be associated with the constitution of the natural bedrock and in the case of where iron, iron oxides and hydroxides appear, they work like a cementing constituent for the quartz particles. The electrical conductivity (EC) values measured for the local soil were very low, ranging between 10.4 and 18.0 μS/cm, indicating that the soil is in its natural state, without ever coming into contact with leachate (highly conductive) from the older existing waste dump.

Electrical Resistivity

In order to assess the changes in the resistivity values with the degree of saturation, contamination and mineralogy of some typical geomaterials found at waste disposal sites, undisturbed soil samples and leachate can be carefully selected and sampled around a particular site, for a controlled and detailed study in the laboratory, in order to interpret the in-situ resistivity maps and profiles obtained during geophysical campaigns. Mondelli et al. (2010a) present laboratory electrical resistivity measurements to detect the contamination plumes surrounding and below a waste deposit. These authors measured electrical resistivity (R) of natural undisturbed samples as well as specimens saturated with distilled water, salt solution and original leachate from a MSW disposal site. While the specimens were naturally air dried during a specific period, the resistivity and the degree of saturation were measured. The system consists of an insulating material (PVC) mold, with two stainless-steel cylindrical electrodes, one on the top and another on the bottom of the specimen for electrical current measurement, similar to the one developed by Daniel (1997). Two more 3 mm diameter stainless-steel electrodes were attached into the middle of the specimen, observing the same distance between them (Wenner array) for differential potential transfer. For data acquisition, a Syscal Pro equipment was used. Figure 16 presents the apparatus used.

Figure. 16: Instrumented undisturbed specimen and Syscal Pro equipment used for laboratory electrical resistivity measurements (Mondelli et al., 2010a).

Mondelli et al. (2010a) concluded that the laboratory test results supported the interpretation of the in-situ test data, identifying the contamination spots, plume depths, capillarity and saturation zones and the different types of local soil. The results demonstrated the high influence of mineralogy and degree of weathering of the tropical soils on electrical resistivity values, also suggesting reference values for the study site. Figure 17 presents the results obtained for samples percolated with leachate (EC = 25,000 µS/cm and R = 0.4 ohm.m). At the beginning of the test, the resistivity values did not exceed 10 ohm.m until a 50 % degree of saturation was achieved. Below this degree of saturation, resistivity increased exponentially. Due to the high electrical conductivity of the leachate, the resistivity values became very low, with no influence of mineralogy in this case. These results could be interpreted using Archie's Law (Archie, 1942), as studied by Mondelli (2008).

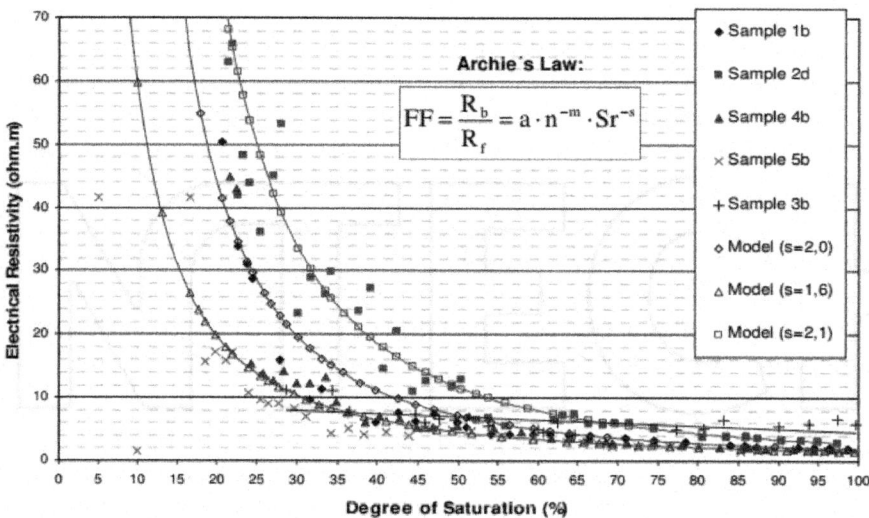

Figure. 17: Electrical resistivity versus degree of saturation for different soil samples percolated with the leachate from a MSW disposal site from Brazil (Mondelli, 2008).

Pollutant Transport Parameters

Estimates of pollutant transport parameters based on laboratory tests such as column, diffusion and batch equilibrium tests are also important in the more detailed phases of geoenvironmental investigations (Freeze & Cherry, 1979; Rowe et al., 1988; Barone et al., 1989; Shackelford & Daniel, 1991; Yong et al., 1992; Shackelford, 1993; ong, 2001). These laboratory tests can be conducted with the objective of a better interpretation of the in-situ test results, and for a numerical modeling performance, using a good conceptual model definition for the problem. Compacted or undisturbed soil samples can be used to assess the pollutant transport parameters, targeting the contamination plume behavior when in contact with the natural environment, using equipment and methods which allows the study of the anisotropy, degree of saturation and chemical extraction. When the determination of pollutant transport parameters is being planned, one of the most important aspects to be defined is the interest pollutants or solutions to be used during the laboratory tests. For geo-environmental investigation of MSW disposal sites, original or artificial leachate can be used. The detection of a concentration of several constituents of leachate (like heavy metals, metals, chloride, organic compounds, chemical oxygen demand and biological oxygen demand), solutions and effluents collected during the tests, need to be chemically analyzed, using emission or atomic absorption spectrometers, flame photometry, chromatography and any other appropriate techniques. The preparation and preservation proceeds to detect any kind of chemical compound need to follow the standards described in APHA, AWWA, WEF (1995).

The laboratory apparatus for column tests also needs to be planned, depending on the sample state (compacted or undisturbed), the hydraulic gradients involved defined in accordance with the field conditions and if the column system will use rigid or flexible wall permeameters. All components of the apparatus were made of non-reactive materials. This test estimates the hydraulic conductivity (k), hydrodynamic dispersion coefficient (D_{hl}) and retardation factor (R_d), due to pollutant solutions or leachate percolation through the soil sample. Once the breakthrough curves (relation between solute concentration and initial concentration – C/C0 in function of the soil pore volumes percolated with the pollutant solution – T) are obtained for each interest solute, considering predominantly advective transport, the hydrodynamic dispersion coefficient (D_{hl}) and the retardation factor (R_d) can be assessed using the following equation proposed by Shackelford (1994):

$$\frac{C}{C_0} = \frac{1}{2}\left\{ erfc\left[\frac{1-(T/R_d)}{2\sqrt{\frac{(t/R_d)}{(uL)/D_h}}} \right] + exp\left[\frac{vL}{D_h} \right] erfc\left[\frac{1+(T/R_d)}{2\sqrt{\frac{(t/R_d)}{(uL)/D_h}}} \right] \right\}$$

(4)

where: t = test time; T = time factor or soil pore volumes percolated with the pollutant solution; u = Darcy'Law specific velocity; L = height of the specimen; (uL/D_h) = Peclet number.

For non-misceble and reactive solutes, the retardation factor (Rd) can be obtained from batch sorption tests. This test can be carried out following ASTM D 4646-03 and ASTM C 1733-10, standards, using a previously defined soil-solution ratio, as 1:4 (25 g of dry soil sieved through a 40 mesh to 100 mL of leachate or pollutant solution). Rd is given by equation 5:

$$R_d = 1 + \frac{\rho_d}{n} \cdot K_d$$

(5)

where: ρ_d = soil specific dried mass; n = soil porosity; K_d = partition coefficient.

Partition coefficient (Kd) represents the angular coefficient of the linear sorption isotherm, defined for an interest solute, during batch sorption tests. For the construction of the sorption isotherms, the amount of soil or solute concentration in the solution can change. When original leachate is used during the test, the solute concentrations remain constant. After a previously determined equilibrium time, the supernatants are filtered, adequatly preserved and stored at 4°C until the chemical analysis, for a limited time, depending on the interest solute. The results are expressed as the adsorption degree, as a function of the concentration. The adsorption degree (S) is defined by the following equation:

$$S = \frac{(C_0 - C_e) \cdot V}{M}$$

(6)

where: C_0 = initial concentration of the solution; C_e = equilibrium concentration; V = volume of the solution in the flask; M = dry soil mass.

Frempong & Yanful (2008) and Mondelli (2008) studied the sorption capacity of tropical soils from Ghana (West Africa) and São Paulo State (Brazil), using original leachate from local MSW disposal sites, and batch and column tests. Both these studies showed the following ion sorption selectivity order: $Ca^{2+} < Cl- < Na^+ \leq Br- < Zn^{2+} < Fe2 < K^+$. A significant finding from these studies is the observation that kaolinite and aluminum and iron oxyhydroxides with variable particle surface charges present in the soil allowed sorption of anions, such as C_1- and B_r- , generally considered conservative nonreactive, based on leachate-liner compatibility studies on soils from temperate regions. Frempong & Yanful (2008) found Rd ranging from 1.1 to 47.9 for Na, K, Br and C_1 for clayey tropical soils. Mondelli (2008) found major values of Rd, ranging from 1 to 383 for N_a, K, Fe and Z_n for sandy tropical soils, with 20 % of clay content. Ritter (1998) classified the solute as essentially immobile when K_d is higher than 10. On the other hand, R_d values display the same order of magnitude as those obtained by Nascentes (2003) with similar soil and range of tested concentrations. Azevedo et al. (2003) point out that the high Rd values obtained by Nascentes (2003) can be the result of the low concentrations of metals used in the pollutant solutions. Figure 19 presents the sorption isotherm obtained for potassium (K) by Mondelli (2008), indicating that adsorption degree (S) increases along with the increase in equilibrium concentration (C_e). In this case, the linear isotherm was the one that best described the results, presenting a better fit to the testing data and K_d value of 29.8 mL/g.

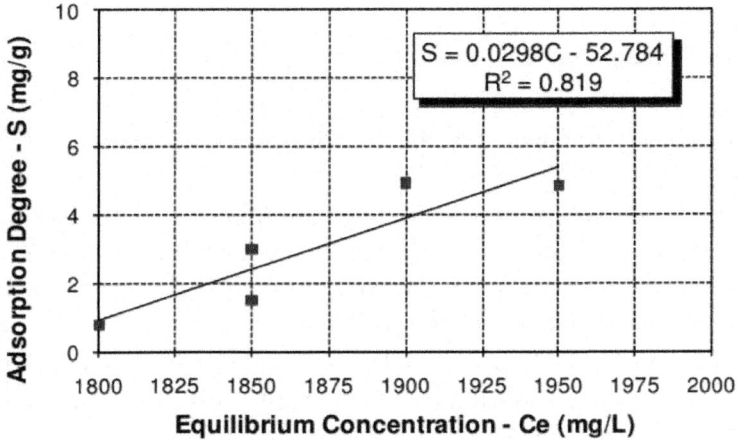

Figure. 18: Linear sorption isotherm obtained for potassium from leachate and a soil sample taken from a MSW disposal site in Brazil (Mondelli, 2008)

CONCLUSION

The chapter shortly presented and discussed the steps of a geo-environmental site investigation program for municipal solid waste disposal sites, and also, different and modern in-situ and laboratory test techniques for this purpose. A literature research was conducted on different areas of knowledge, comprising the current problems on the inadequate disposal of waste, environmental liabilities and techniques for the investigation of contaminated sites. Research results on Environmental Geotechnics were also included, aimed at the appropriate monitoring and proper installation of sanitary landfills. Therefore, it can be concluded that it is fundamental to know exactly all the goals, local interests, background factors and resources involved in each case, for an effective and optimized geoenvironmental site investigation. This approach will provide positive results for the local population and for the physical environment.

ACKNOWLEDGMENTS

The authors gratefully acknowledge the Brazilian research agencies FAPESP (State of São Paulo Research Foundation), CAPES (Brazilian Federal Agency for Support and Evaluation of Graduate Education) and CNPq (National Council for Scientific and Technological Development) for funding their researches.

REFERENCES

1. ABNT NBR – 13600 (1996). Solo – Determinação do Teor de Matéria Orgânica por Queima a 440°C. Brazilian Association for Technical Standards, São Paulo-SP, Brazil.

2. APHA, AWWA, WEF (1995). Standard Methods for the Examination of Water and Wastewater. American Public Health Association, American Water Works Association, Water Environmental Federation. Managing Editor Mary Ann H. Franson. 19th Edition, U. S. A., pp. 3-1 – 3-2.

3. ASTM D 5778-07. Standard Test Method for Electronic Friction Cone and Piezocone Penetration Testing of Soils. ASTM International, Pennsylvania, U.S.A.

4. ASTM D 4646- 03 (Reapproved 2008). Standard Test Method for 24-h Batch-Type Measurement of Contaminant Sorption by Soils and Sediments. ASTM Intern., Pennsylvania, U.S.A.

5. ASTM C 1733-10. Standard Test Method for Distribution Coefficients of Inorganic Species by the Batch Method. ASTM Intern., Pennsylvania, U.S.A.

6. Anirban, D. E.; Matasovic, N. & Dunn, R. J. (2004). Site Characterization of Five Hazardous Waste Landfills. Proceedings of the 2nd International Conference on Site Characterization, Porto, Portugal, September 2004, pp. 1075-1080.

7. Archie, G. E. (1942). The Electrical Resistivity Log as an Aid in Determining Some Reservoir Characteristics. Transactions of American Institute of Mining and Metallurgy Engineering, Vol.146, pp. 54-62.

8. Azevedo, I. C. D.; Nascentes, C. R.; Matos, A. T. & Azevedo, R. F. (2006). Determination of Transport Parameters for Heavy Metal in Residual Compacted Soil Using Two Methodologies. Canadian Journal of Civil Engineering, Vol.33, pp. 912-917.

9. Azevedo, I. C. D. , Nascentes, C. R.; Azevedo, R. F.; Matos, A. T. & Guimarães, L. M. (2003). Coeficiente de Dispersão Hidrodinâmica e Fator de Retardamento de Metais Pesados em Solo Residual Compactado. Solos e Rochas, Vol.26, pp.229-249.

10. Barone, F. S.; Yanful, E. K.; Quigley, R. M. & Rowe, R. K. (1989). Effect

of Multiple Contaminant Migration on Diffusion and Adsorption of Some Domestic Waste Contaminants in a Natural Clayey Soil. Canadian Geotechnical Journal, Vol.26, pp. 189-198.

11. Bolinelli Jr., H. L. (2004). Piezocone de Resistividade: Primeiros Resultados de Investigação Geoambiental em Solos Tropicais. MSc, Department of Civil Engineering, São Paulo State University, Bauru-SP, Brazil.

12. Boscov, M. E. G.; Cunha, I. I. & Saito, R. (2001). Radium Migration Through Clay Liners at Waste Disposal Sites. Science of t he Total Environment, Vol.266, pp. 259-264.

13. Campanella, R. G. (2008). Geo-Environmental Site Characterization. Proceedings of the 3rd International Conference on Geotechnical and Geophysical Site Characterization, Taipei –Taiwan, Taylor & Francis Group, pp. 3-16.

14. CETESB (2005). Relatório de Estabelecimento de Valores Orientadores para Solos e ÁguasSubterâneas no Estado de São Paulo. São Paulo-SP, Brasil.

15. Cherry, J. A.; Gillham, R. W.; Anderson, E. G. & Johnson, P. E. (1983). Migration of Contaminants in Groundwater at a Landfill: a Case Study 2. Groundwater Monitoring Devices. Journal of Hydrology, Vol.63, pp. 31-49.

16. Christensen, T. H., Bjerg, P. L. & Kjeldsen, P. (2000). Natural attenuation: a feasible approach to remediation of ground water pollution at landfills? Ground Water Monitoring & Remediation, Vol.20, pp. 69-77.

17. Dahlin, T.; Bernstone, C. & Loke, M. H. (2002). A 3-D Resistivity Investigation of a Contaminated Site at Lernacken, Sweden. Geophysics, Vol.67, pp. 1692-1700.

18. Daniel, C.R. (1997). An Investigation of the Factors Affecting Bulk Soil Electrical Resistivity. BASc, University of British Columbia, Department of Civil Engineering. Davies, M. P. & Campanella, R. G. (1995). Environmental Site Characterization Using In-Situ Testing Methods, 48th Canadian Geotechnical Conference, Vancouver-BC, Canada.

19. Elis, V. R. & Zuquette, L. V. (2002). Caracterização Geofísica de Áreas Utilizadas para Disposição de Resíduos Sólidos Urbanos. Revista Brasileira de Geociências, Vol.32, pp. 119-134.

20. Frempong, E. M. & Yanful, E. K. (2008). Interactions Between Three Tropical Soils and Municipal Solid Waste Landfill Leachate. Journal of Geotechnical and Geoenvironmental Engineering, Vol.134, pp.379-396.

21. Freeze, R. A. & Cherry, J. A. (1979). Groundwater. Prentice Hall, Inc. Englewood Cliffs, New Jersey, U. S. A.

22. Geoprobe Direct Image (2004). MP 6500 – Membrane Interface Probe (MIP): User Manual. Document N. 3083, Revision 1.0 10-12-2004, Kejr, Inc., Salina, Kansas, U.S.A.

23. Grazinolli, P. L.; Costa, A.; Campos, T. M. P. & Vargas Jr., E. A. (1999). Aplicações do Radar de Penetração no Solo (GPR) e da Eletrorresistividade para a Detecção de Compostos Orgânicos. Proceeding of the 4º Congresso Brasileiro de Geotecnia Ambiental, São José dos Campos, Brazil, December 1999, pp 127-13.

24. Innov-X Systems Alpha Series (2007). X-Ray Fluorescence Spectrometers. P/N 100392, Revision B. Woburn, MA, U. S. A.

25. IPT (2011). Estudo de Viabilidade para Ampliação da Área de Disposição Final de Resíduos Sólidos Urbanos no Município de Anhembi-SP. Institute for Technological Research, Center for Energy and Environmental Technologies. Parecer Técnico 19455-301/11, complementar ao Parecer Técnico 19287-301/11.

26. Kjeldsen, P.; Bjerg, P. L.; Rügge, K.; Christensen, T. H. & Pedersen, J. K. (1998). Characterization of an Old Municipal Landfill (Grindsted, Denmark) as a Groundwater Pollution Source: Landfill Hydrology and Leachate Migration. Waste Management & Research, Vol.16, pp. 14-22.

27. Lago, A. L.; Silva, E. M. A.; Elis, V. R. & Giacheti, H. L. (2003). Aplicação de Eletrorresistividade e Polarização Induzida em Área de Disposição de Resíduos Sólidos Urbanos em Bauru-SP. Proceedings of the 8th International Congress of the Brazilian Geophysical Society, Rio de Janeiro, Brazil, November 2003, CD-ROM.

28. Lambe, T. W. & Whitman, R. V. (1969). Soil Mechanics. New York: J. Wiley, 553p. Lan, T. N. (1977). Um Nouvel Esai D'identification des Sols: L'essai au Bleu de Methyléne. Bulletin Liaison Laboratoire dês Ponts et Chaussée, Paris, pp.136-137.

29. Larsson, R. (1995). Use of a Thin Slot as Filter in Piezocone Tests. Proceedings of International Symposium on Cone Penetration Testing (CPT'95), Linköping, Sweden, pp. 35–40.

30. Leite, A. L. & Paraguassú, A. B. (2002). Diffusion of Inorganic Chemicals in Some Compacted Tropical Soils. Proceedings of the 4th International Congress In Environmental Geotechnics, Rio de Janeiro, Brazil, August 2002, pp 39-45.

31. Loke, M. H. & Barker, R. D. (1996). Practical Techniques for 3D Resistivity Survey and Data Inversion. Geophysical Prospecting, Vol.44,

pp. 499-523.

32.   Lunne, T.; Robertson, P. K. & Powell, J. (1997). Cone Penetration Test in Geotechnical Practice. Blackie Academic & Professional, London.

33.   Mackay, D. M.; Freyberg, D. L.; Roberts, P. V. & Cherry, J. A. (1986). A Natural Gradient Experiment on Solute Transport in a Sand Aquifer: 1. Approach and Overview of Plume Movement. Water Resources Research, Vol.22, pp. 2017-2029.

34.   MacFarlane, D. S.; Cherry, J. A.; Ghillham, R. W. & Sudicky, E. A. (1983). Migration of Contaminants in Groundwater at a Landfill: a Case Study, 1. Groundwater Flow and Plume Delineation. Journal of Hydrology, Vol.63, pp. 1-29.

35.   McNeill, J. D. (1980). Electromagnetic Terrain Conductivity Measurement at Low Induction Numbers. Geonics Technical Note, No.6, 15p.

36.   Mondelli, G. (2008). Integração de Diferentes Técnicas de Investigação para Avaliação da Poluição e Contaminação de uma Área de Disposição de Resíduos Sólidos Urbanos. Ph.D. Thesis, Department of Geotechnical Engineering, University of São Paulo, São Carlos-SP, Brazil.

37.   Mondelli, G.; Giacheti, H. L.; Boscov, M. E. G.; Elis, V. R. & Hamada, J. (2007). Geoenvironmental Site Investigation Using Different Techniques in a Municipal Solid Waste Disposal Site in Brazil. Environmental Geology, Vol.52, pp. 871-887.

38.   Mondelli, G.; Giacheti, H. L. & Elis, V. R. (2010a). The Use of Resistivity for Detecting MSW Contamination Plumes in a Tropical Soil Site. Proceedings of 6th International Conference on Environmental Geotechnics, New Delhi, India, pp. 1544-1549.

39.   Mondelli, G.; Giacheti, H. L. & Howie, J. A. (2010b). Interpretation of Resistivity Piezocone Tests in a Contaminated Municipal Solid Waste Disposal Site. Geotechnical Testing Journal, Vol. 33, N. 2, pp. 123-136.

40.   Monteiro Santos, F. A. (2004). 1-D Laterally Constrained Inversion of EM34 Profiling Data. Journal of Applied Geophysics, Vol.56, pp. 123-134.

41.   Nascentes, C. R. (2003). Coeficiente de Dispersão Hidrodinâmica e Fator de Retardamento de Metais Pesados em Solo Residual Compactado. MSc, Department of Civil Engineering, Federal University of Viçosa, Viçosa-MG, Brazil.

42.   Pejon, O. J. (1992). Mapeamento Geotécnico da Folha de Piracicaba-SP (Escala 1:100.000): Estudo de Aspectos Metodológicos, de Caracterização e de Apresentação dos Atributos. Ph.D.

43. Thesis, Department of Geotechnical Engineering, University of São Paulo, São Carlos-SP, Brazil.

44. Porsani, J. L.; Malagutti Filho, W.; Elis, V. R.; Fisseha, S., Dourado, J. C. & Moura, H. P. (2004). The Use of GPR and VES in Delineating a Contamination Plume in a Landfill Site: A Case Study in SE Brazil. Journal of Applied Geophysics, Vol.55, pp.199-209.

45. Ritter, E. (1998). Efeito da Salinidade na Difusão e Sorção de Alguns Íons Inorgânicos em um Solo Argiloso Saturado. PhD, Department of Civil Engineering, Federal University of Rio de Janeiro, Rio de Janeiro-RJ, Brazil.

46. Robertson, P. K. (1998). Geo-Environmental Investigation, Characterization and Monitoring Using Penetration Techniques. Simpósio Brasileiro de Geotecnia Ambiental, São PauloSP, Brazil.

47. Robertson, P. K.; Campanella, R. G., Gillespie, D. & Greig, J. (1986). Use of piezometer cone data, Proc. of the In-Situ-86, ASCE Specialty Conference, p. 1263-1280

48. Robertson, P. K. & Cabal, K. L. (2008). Guide to Cone Penetration Testing for Geo-Environmental Engineering. Gregg Drilling & Testing, Inc., 2nd Edition, 84 p.

49. Robertson, P. K. & Campanella, R. G. (1988). Guidelines for Geotechnical Design Using CPT and CPTU Data. Report FHWA, 340p.

50. Rowe, R. K.; Caers, C. J. & Barone, F. S. (1988). Laboratory Determination of Diffusion and Distribution Coefficients of Contaminants Using Undisturbed Clayey Soil. Canadian Geotechnical Journal, Vol.25, No.1, pp. 108-118.

51. Shackelford, C. D. (1993). Contaminant Transport. Geotechnical Practice for Waste Disposal, Chapman & Hall, London, pp. 33-65.

52. Shackelford, C. D. (1994). Critical Concepts for Column Testing. Journal of Geotechnical Engineering, Vol.120, pp.1804-1828.

53. Shackelford, C. D. & Daniel, D. E. (1991). Diffusion in Saturated Soil. I: Background. Journal of Geotechnical Engineering, Vol.117, No.3, pp. 467-484.

54. Shinn II, J.D. & Bratton, W.L. (1995). Innovations with CPT for environmental site characterization», Proceedings of CPT›95, Vol. 2, pp. 93-98.

55. Stuermer, M. M. (2005). Contribuição ao Estudo de um Solo Saprolítico como Revestimento Impermeabilizante de Fundo de Aterros de Resíduos. PhD, Department of Structures and Geotechnical Engineering, University

of São Paulo, São Paulo-SP, Brazil.

56. Telford, W. M.; Geldart, L. P.; Sheriff, R. E. & Keys, D. A. (1990). Applied Geophysics, Cambridge University Press, 860 p.

57. Thornthwaite, C.W. & Mather, J.R. (1955). The water balance. Publications in Climatology. New Jersey: Drexel Institute of Technology, 104p.

58. US EPA (1989). Seminar on Site Characterization for Subsurface Remediations. United States Environmental Protection Agency, Technology Transfer, Report CERI-89-224, 350p.

59. US EPA (1993). SW-846 pH in Liquid and Soil. Method 9040 (Liquid) and SW-846 Method 9045 (Soil).

60. Ustra, A. T.; Elis, V. R.; Mondelli, G.; Zuquette, L. V. & Giacheti, H. L. (2011). Case Study: A 3D Resistivity and Induced Polarization Imaging from Downstream a Waste Disposal Site in Brazil. Environmental Earth Sciences, in press (online).

61. Weemes, I. (1990). A Resistivity Cone Penetrometer for Ground-Water Studies. MASc, University of British Columbia, Department of Civil Engineering.

62. Yong, R. N. (2001). Geoenvironmental Engineering: Contaminated Soils, Pollutant Fate & Mitigation. CRC Press, USA, 307p

63. Yong, R. N.; Mohamed, A. M. O. & Warkentin, B. P. (1992). Principles of Contaminant Transport in Soils. Elsevier Science Publishers B.V., 327 p.

64. Zuquette, L. V.; Palma, J. B. & Pejon, O. J. (2005). Environmental Assessment of anUncontrolled Sanitary Landfill. Bulletin of Engineering Geology and the Environment, Vol.64, pp.257-271.

CITATION

CHAPTER 1

Paulo Cesar Fernandes da Silva and John Canning Cripps (2011). Geo-environmental Terrain Assessments Based on Remote Sensing Tools: A Review of Applications to Hazard Mapping and Control, Environmental Management in Practice, Dr. Elzbieta Broniewicz (Ed.), ISBN: 978-953-307-358-3, InTech, DOI: 10.5772/24487.

CHAPTER 2

B. He, J. Chen, C. Chen and Y. Liu, "Mineral Prospectivity Mapping Method Integrating Multi-Sources Geology Spatial Data Sets and Case-Based Reasoning," Journal of Geographic Information System, Vol. 4 No. 2, 2012, pp. 77-85. doi: 10.4236/jgis.2012.42011.

CHAPTER 3

S. Singh and P. Singh, "Mapping Spatial Data on the Web Using Free and Open-Source Tools: A Prototype Implementation," Journal of Geographic Information System, Vol. 6 No. 1, 2014, pp. 30-39. doi: 10.4236/jgis.2014.61004.

CHAPTER 4

A. Suhardiman, S. Tsuyuki, M. Sumaryono and Y. Sulistioadi, "Geostatistical Approach for Site Suitability Mapping of Degraded Mangrove Forest in the Mahakam Delta, Indonesia," Journal of Geographic Information System, Vol. 5 No. 5, 2013, pp. 419-428. doi: 10.4236/jgis.2013.55040.

CHAPTER 5

Emmanuel Vassilakis, Remote Sensing of Environmental Change in the Antirio Deltaic Fan Region, Western Greece, doi:10.3390/rs2112547.

CHAPTER 6

Issaak Parcharidis Sotiris Kokkalas, Ioannis Fountoulis and Michael Foumelis, Detection and Monitoring of Active Faults in Urban Environments: Time Series Interferometry on the Cities of Patras and Pyrgos (Peloponnese, Greece), doi:10.3390/rs1040676.

CHAPTER 7

Slavoljub Dragicevic, Nenad Zivkovic, Mirjana Roksandic, Stanimir Kostadinov, Ivan Novkovic, Radislav Tosic, Milomir Stepic, Marija Dragicevic and Borislava Blagojevic (2012). Land Use Changes and Environmental Problems Caused by Bank Erosion: A Case Study of the Kolubara River Basin in Serbia, Environmental Land Use Planning, Dr. Seth Appiah-Opoku (Ed.), ISBN: 978-953-51-0832-0, InTech, DOI: 10.5772/50580.

CHAPTER 8

Seth Appiah-Opoku and Crystal Taylor (2012). Environmental Land Use and the Ecological Footprint of Higher Learning, Environmental Land Use Planning, Dr. Seth Appiah-Opoku (Ed.), ISBN: 978-953-51-0832-0, InTech, DOI: 10.5772/48191.

CHAPTER 9

Dirk Loehr (2012). The Role of Tradable Planning Permits in Environmental Land Use Planning: A Stocktake of the German Discussion, Environmental Land Use Planning, Dr. Seth Appiah-Opoku (Ed.), ISBN: 978-953-51-0832-0, InTech, DOI: 10.5772/50469.

CHAPTER 10

A. Gharagozlou, H. Nazari and M. Seddighi, "Spatial Analysis for Flood Control by Using Environmental Modeling," Journal of Geographic Information System, Vol. 3 No. 4, 2011, pp. 367-372. doi: 10.4236/jgis.2011.34035.

CHAPTER 11

Gebreyesus Brhane Tesfahunegn, Lulseged Tamene, and Paul L. G. Vlek, "Soil Erosion Prediction Using Morgan-Morgan-Finney Model in a GIS Environment in Northern Ethiopia Catchment," Applied and Environmental Soil Science, vol. 2014, Article ID 468751, 15 pages, 2014. doi:10.1155/2014/468751.

CHAPTER 12

Giulliana Mondelli, Heraldo Luiz Giacheti and Vagner Roberto Elis (2012). Geo-Environmental Site Investigation for Municipal Solid Waste Disposal Sites, Municipal and Industrial Waste Disposal, Dr. Xiao-Ying Yu (Ed.), ISBN: 978-953-51-0501-5, InTech, DOI: 10.5772/28835.

INDEX

A

Analytic Hierarchy Process (AHP) 9
Application program interfaces (APIs)
    60
Artificial electrical 243

C

Cadastral maps 142, 148
case-based reasoning (CBR) 1, 2
Common hazards 202
common language runtime (CLR) 60
Crustal-scale 5, 7

D

Deactivated dump 239
Digital elevation model (DEM) 79, 215

E

Earthquake cycle 116
Earth Resources Observation and Science (EROS) 102
Ecological aspects 151
Ecological footprint 161, 162, 164, 166,
    167, 168, 170, 172, 173, 174, 175,
    177, 179, 181
Economic consequences 151
Environmental agencies 239, 241
Environmental changes 99, 101, 111
Environmental modeling 204
Experiment regarding 10

F

Flood hazard 201, 203, 205

G

Geodynamic phenomena 7, 11, 12, 19
Geo-environmental 1, 2, 3, 5, 6, 15, 26,
    39
geographical information system (GIS)
    78
Geographical Information System (GIS)
    211
geographic information systems (GIS)
    57
Geological hazards 1
Geological structures 11, 14, 17, 34
Geomorphological analysis 139
Geophysical methods 242
global warming potentials (GWPs) 188

I

Integrated development environment
    (IDE) 60, 61
Interferometric Point Target Analysis
    (IPTA) 116

K

Kephalonia Transform Fault (KTF) 117

L

Last decade 111

M

marginal abatement costs (MACA+B)
    189
Massive degradation 78
Morgan-Morgan-Finney (MMF) 211,
    213

O

Object-relational database management
    system (ORDBMS) 62
Operational guidance 77

P

Persistent Scatterers Interferometry (PSI)
    116
Physical data 79
Planning system 185, 191, 192

R

Regime sketched 193
Resistivity Method 243, 244
Ridgecrest East 164, 165, 166, 167, 168,
    169, 170, 171, 172, 173, 174, 175
Rock resistivity 244

S

Sandy-clay 152
Similarity ranged 5
Single Look Complex (SLC) 125
Spatial information 201
Statistical approaches 16
Statistics and sampling 203
Surface layer-based 83

T

Technical problems 167
Terrain susceptibility 6
Thematic Mapper sensor (TM) 102
Tradable planning 183, 186, 190, 191,
    192, 196

U

Universal Soil Loss Equation (USLE)
    213
Universal Transverse Mercator (UTM)
    102
U.S. Geological Survey (USGS) 102

V

Vertical electrical soundings (VES) 245

W

weights-of-evidence (WOE) 1, 2